FE/EIT EXAM PREPARATION

CIVIL ENGINEERING

Donald G. Newnan, PhD, PE Civil Eng., Editor

Alan Williams, PhD, SE, Chartered Eng., California Department of Transportation

Bruce Larock, PhD, PE, University of California, Davis

Kenneth Williamson, PhD, PE, Oregon State University

Braja Das, PhD, PE, California State University, Sacramento

Robert W. Stokes, PhD, Kansas State University

Michael Taylor, PhD, University of California, Davis

Lincoln D. Jones, PE, San Jose State University

D1265256

President: Roy Lipner
Publisher: Evan M. Butterfield
Senior Development Editor: Laurie McGuire
Managing Editor, Production: Daniel Frey
Quality Assurance Editor: David Shaw
Creative Director: Lucy Jenkins

Published by Kaplan® AEC Education,
a division of Dearborn Financial Publishing, Inc.®,
a Kaplan Professional Company®

30 South Wacker Drive
Chicago, IL 60606-7481
(312) 836-4400
www.engineeringpress.com

CONTENTS

CHAPTER 9

Legal and Professional Aspects: Ethics

APPENDIX A

Afternoon Sample Exam   125

BECOMING A PROFESSIONAL ENGINEER

To achieve registration as a Professional Engineer, there are four distinct steps: (1) education, (2) the Fundamentals of Engineering/Engineer-in-Training (FE/EIT) exam, (3) professional experience, and (4) the professional engineer (PE) exam. These steps are described in the following sections.

Education

Generally, no college degree is required to be eligible to take the FE/EIT exam. The exact rules vary, but all states allow engineering students to take the FE/EIT exam before they graduate, usually in their senior year. Some states, in fact, have no education requirement at all. One merely need apply and pay the application fee. Perhaps the best time to take the exam is immediately following completion of related coursework. For most engineering students, this will be the end of the senior year.

Fundamentals of Engineering/ Engineer-In-Training Examination

This eight-hour, multiple-choice examination is known by a variety of names— Fundamentals of Engineering, Engineer-in-Training (EIT), and Intern Engineer— but no matter what it is called, the exam is the same in all states. It is prepared and graded by the National Council of Examiners for Engineering and Surveying (NCEES).

Experience

States that allow engineering seniors to take the FE/EIT exam have no experience requirement. These same states, however, generally will allow other applicants to substitute acceptable experience for coursework. Still other states may allow a candidate to take the FE/EIT exam without any education or experience requirements.

Typically, four years of acceptable experience is required before one can take the Professional Engineer exam, but the requirement may vary from state to state.

Professional Engineer Examination

The second national exam is called Principles and Practice of Engineering by NCEES, but many refer to it as the Professional Engineer exam or PE exam. All states, plus Guam, the District of Columbia, and Puerto Rico, use the same NCEES exam. Review materials for this exam are found in other engineering license review books.

FUNDAMENTALS OF ENGINEERING/ ENGINEER-IN-TRAINING EXAMINATION

Laws have been passed that regulate the practice of engineering in order to protect the public from incompetent practitioners. Beginning in 1907 the individual states began passing *title* acts regulating who could call themselves engineers and offer services to the public. As the laws were strengthened, the practice of engineering was limited to those who were registered engineers, or to those working under the supervision of a registered engineer. Originally the laws were limited to civil engineering, but over time they have evolved so that the titles, and some-times the practice, of most branches of engineering are included.

There is no national licensure law; licensure is based on individual state laws and is administered by boards of registration in each state. Table 1.1 is a listing of the state boards of registration and territory.

Table 1.1 State Boards of Registration for Engineers

State	Web site	Telephone
AL	www.bels.state.al.us	334-242-5568
AK	www.dced.state.ak.us/occ/pael.htm	907-465-1676
AZ	www.btr.state.za.us	602-364-4930
AR	www.state.ar.us/pels	501-682-2824
CA	Dca.ca.gov/pels/contacts/htm	916-263-2230
CO	Dova.state.co.us/engineers_surveyors	303-894-7788
CT	State.ct.us/dcp	860-713-6145
DE	www.dape.org	302-368-6708
DC		202-442-4320
FL	www.fbpe.org	850-521-0500
GA	www.sos.state.ga.us/plb/pels/	478-207-1450
GU	www.guam-peals.org	671-646-3138
HI	www.Hawaii.gov/dcca/pbl	808-586-2702
ID	www.state.id.us/ipels/index.htm	208-334-3860
IL	www.kpr.state.il.us	217-785-0877
IN	www.in.gov/pla/bandc/engineers	317-232-2980
IA	www.ia.us/government/com	515-281-4126
KS	www.accesskansas.org/ksbtp	785-296-3053
KY	www.kyboels.state.ky.us	502-573-2680
LA	www.lapels.com	225-925-6291
ME	www.professionals.maineusa.com	207-287-3236
MD	www.dllr.state.md/us	410-230-6322
MA	www.state.ma.us/reg	617-727-9957
MI	www.Michigan.gov/cis/0,1607, 7-154-10557_12992_14016___00.html	517-241-9253
MN	www.aelslagid.state.mn.us	651-296-2388
MS	www.pepls.state.ms.us	601-359-6160
MO	www.pr.mo.gov/apelsla.asp	573-751-0047
MP		(011) 670-234-5897
MT	www.discoveringmontana.com/dli/bsd/license/bsd.board/pel_board/board_page.htm	406-841-2367
NE	www.ea.state.ne.us	402-471-2021
NV	www.boe.state.nv.us	775-688-1231
NH	www.state.nh.us/jtboard/home.htm	603-271-2219
NJ	www.state.nj.us	973-504-6460
NM	www.state.nm.us pepsboard	505-827-7561
NY	www.op.nysed.gov	518-474-3817 x140
NC	www.ncbels.org	919-881-4000
ND	www.ndpelsboard.org/	701-258-0786
OH	www.ohiopeps.org	614-466-3651
OK	www.pels.state.ok.us/	405-521-2874
OR	www.osbeels.org	503-362-2666
PA	www.dos.state.pa.us/eng	717-783-7049

(*Continued*)

Table 1.1 State Boards of Registration for Engineers (*Continued*)

State	Web site	Telephone
PR	P.O. Box 3271, San Juan 00904	787-722-2122
RI	www.bdp.state.ri.us	401-222-2565
SC	www.llr.state.sc.us/POL/Engineers	803-896-4422
SD	www.state.sd.us/dol/boards/engineer	605-394-2510
TN	www.state.tn.us/commerce/boards/ae/	615-741-3221
TX	www.tbpe.state.tx.us	512-440-7723
UT	www.dopl.utah.gov	801-530-6632
VT	vtprofessionals.org	802-828-3256
VI	www.dlca.gov.vi/pro-aels.html	340-773-2226
VA	www.state.va.us/dpor	804-367-8514
WA	www.dol.wa.gov/engineers/engfront.htm	360-664-1595
WV	www.wvpebd.org	304-558-3554
WI	www.drl.state.wi.us	608-261-7096
WY	www.wrds.uwyo.edu/wrds/borpe/borpe.html	307-777-6155

Examination Structure

The FE/EIT exam is divided into a morning four-hour section and an afternoon four-hour section. There are 120 questions in the morning section and 60 in the afternoon.

The morning exam covers the topics that make up roughly the first $2\frac{1}{2}$ years of a typical engineering undergraduate program. All examinees take the same morning exam.

Seven different exams are in the afternoon test booklet, one for each of the following six branches: civil, mechanical, electrical, chemical, industrial, environmental. A general exam is included for those examinees not covered by the six engineering branches. Each of the six branch exams consists of 60 problems covering coursework in the specific branch of engineering. The general exam, also 60 problems, has topics that are similar to the morning topics. Thus the afternoon exam may benefit those specializing in one of the six engineering branches. Most of the test's topics cover the third and fourth year of college courses. These are the courses you will use for the balance of your engineering career, so the test becomes focused to your own needs. Graduate engineers will find the afternoon branch test to their advantage, as the broad fundamentals test usually causes them to do a good deal of review of their earliest classwork.

We recommend that civil, mechanical, electrical, chemical, enviromental, and industrial engineers take their branch exam. All others should take the general examination. Analysis of pass rates on previous exams shows that examinees taking the discipline-specific exam in their branch of engineering typically do better than those taking the general exam!

At the beginning of the afternoon test period, examinees will mark the answer sheet as to which branch exam they are taking. You could quickly scan the test, judge the degree of difficulty of the general versus the branch exam, then choose the test to answer. We do not recommend this practice, as you would waste time in determining which test to write. Further, you could lose confidence during this indecisive period.

Table 1.2 summarizes the major subjects for the six exams, including the percentage of problems you can expect to see on each one.

Table 1.2 FE Afternoon Exams

Branch Exam	Topics	Percentage of Problems
Chemical	Chemical reaction engineering	10
	Chemical thermodynamics	10
	Computer and numerical methods	5
	Heat transfer	10
	Mass transfer	10
	Material/energy balances	15
	Pollution prevention	5
	Process control	5
	Process design and economics evaluation	10
	Process equipment design	5
	Process safety	5
	Transport phenomena	10
Civil	Computers and numerical methods	10
	Construction management	5
	Environmental engineering	10
	Hydraulics and hydrologic systems	10
	Legal and professional aspects	5
	Soil mechanics and foundations	10
	Structural analysis	10
	Structural design	10
	Surveying	10
	Transportation facilities	10
	Water purification and treatment	10
Electrical	Analog electronic circuits	10
	Communications theory	10
	Computer and numerical methods	5
	Computer hardware engineering	5
	Computer software engineering	5
	Control systems theory and analysis	10
	Digital systems	10
	Electromagnetic theory and applications	10
	Instrumentation	5
	Network analysis	10
	Power systems	5
	Signal processing	5
	Solid state electronics and devices	10
Environmental	Water resources	25
	Water and wastewater engineering	30
	Air-quality engineering	15
	Solid-and hazardous-waste engineering	15
	Environmental science and management	15
General	Chemistry	7.5
	Computers	5
	Dynamics	7.5

(Continued)

Table 1.2 FE Afternoon Exams (*Continued*)

Branch Exam	Topics	Percentage of Problems
	Electrical circuits	10
	Engineering economics	5
	Ethics	5
	Fluid mechanics	7.5
	Materials science/structure of matter	5
	Mathematics	20
	Mechanics of materials	7.5
	Statics	10
	Thermodynamics	10
Industrial	Computer computations and modeling	5
	Design of industrial experiments	5
	Engineering economics	5
	Engineering statistics	5
	Facility design and location	5
	Industrial cost analysis	5
	Industrial ergonomics	5
	Industrial management	5
	Information system design	5
	Manufacturing process	5
	Manufacturing systems design	5
	Material handling system design	5
	Mathematical optimization and modeling	5
	Production planning and scheduling	5
	Productivity measurement and management	5
	Queuing theory and modeling	5
	Simulation	5
	Statistical quality control	5
	Total quality management	5
	Work performance and methods	5
Mechanical	Automatic controls	5
	Computer	5
	Dynamic systems	10
	Energy conversion and power plants	5
	Fans, pumps, and compressors	5
	Fluid mechanics	10
	Heat transfer	10
	Material behavior/processing	5
	Measurement and instrumentation	10
	Mechanical design	10
	Refrigeration and HVAC	5
	Stress analysis	10
	Thermodynamics	10

Examination Procedure

Before the morning four-hour session begins, the proctors pass out exam booklets and a scoring sheet to each examinee. Space is provided on each page of the examination booklet for scratchwork. The scratchwork will *not* be considered in the scoring. Proctors will also provide each examinee with a mechanical pencil for use in recording answers; this is the only writing instrument allowed. Do not bring your own lead or eraser. If you need an additional pencil during the exam, a proctor will supply one.

The examination is closed book. You may not bring any reference materials with you to the exam. To replace your own materials, NCEES has prepared a *Fundamentals of Engineering (FE) Supplied-Reference Handbook*. The handbook contains engineering, scientific, and mathematical formulas and tables for use in the examination. Examinees will receive the handbook from their state registration board prior to the examination. The *FE Supplied-Reference Handbook* is also included in the exam materials distributed at the beginning of each four-hour exam period.

There are three versions (A, B, and C) of the exam. These have the major subjects presented in a different order to reduce the possibility of examinees copying from one another. The first subject on your exam, for example, might be fluid mechanics, while the exam of the person next to you may have electrical circuits as the first subject.

The afternoon session begins following a one-hour lunch break. The afternoon exam booklets will be distributed along with a scoring sheet. There will be 60 multiple choice questions, each of which carries twice the grading weight of the morning exam questions.

If you answer all questions more than 30 minutes early, you may turn in the exam materials and leave. If you finish in the last 30 minutes, however, you must remain to the end of the exam period to ensure a quiet environment for all those still working, and to ensure an orderly collection of materials.

Examination-Taking Suggestions

Those familiar with the psychology of examinations have several suggestions for examinees:

1. There are really two skills that examinees can develop and sharpen. One is the skill of illustrating one's knowledge. The other is the skill of familiarization with examination structure and procedure. The first can be enhanced by a systematic review of the subject matter. The second, exam-taking skills, can be improved by practice with sample problems—that is, problems that are presented in the exam format with similar content and level of difficulty.

2. Examinees should answer every problem, even if it is necessary to guess. There is no penalty for guessing. The best approach to guessing is to first eliminate the one or two obviously incorrect answers among the four alternatives. If this can be done, the chance of selecting a correct answer obviously improves from 1 in 4 to 1 in 2 or 3.

3. Plan ahead with a strategy and a time allocation. There are 120 morning problems in 12 subject areas. Compute how much time you will allow for each of the 12 subject areas. You might allocate a little less time per problem for the areas in which you are most proficient, leaving a little more time in subjects that are more difficult for you. Your time plan should include

a reserve block for especially difficult problems, for checking your scoring sheet, and finally for making last-minute guesses on problems you did not work. A time plan gives you the confidence of being in control. Misallocation of time for the exam can be a serious mistake. Your strategy might also include time allotments for two passes through the exam—the first to work all problems for which answers are obvious to you, the second to return to the more complex, time-consuming problems and the ones at which you might need to guess.

4. Read all four multiple-choice answers options before making a selection. All distractors (wrong answers) are designed to be plausible. Only one option will be the best answer.

5. Do not change an answer unless you are absolutely certain you have made a mistake. Your first reaction is likely to be correct.

6. If time permits, check your work.

7. Do not sit next to a friend, a window, or other potential distraction.

License Review Books

To prepare for the FE/EIT exam you need two or three review books.

1. *Fundamentals of Engineering: FE Exam Preparation* to provide a review of the common four-hour morning examination.

2. An afternoon supplement book if you are going to take one of the six branch exams.

3. *Fundamentals of Engineering (FE) Supplied-Reference Handbook*. At some point this NCEES-prepared book will be provided to applicants by their State Registration Board. You may want to obtain a copy sooner so you will have ample time to study it before the exam. You must, however, pay close attention to the *FE Supplied-Reference Handbook* and the notation used in it, because it is the only book you will have at the exam.

Textbooks

If you still have your university textbooks, they can be useful in preparing for the exam, unless they are out of date. To a great extent the books will be like old friends with familiar notation. You probably need both textbooks and license review books for efficient study and review.

Examination Day Preparations

The exam day will be a stressful and tiring one. You should take steps to eliminate the possibility of unpleasant surprises. If at all possible, visit the examination site ahead of time. Try too determine such items as

1. How much time should I allow for travel to the exam on that day? Plan to arrive about 15 minutes early. That way you will have ample time, but not too much time. Arriving too early, and mingling with others who are also anxious, can increase your anxiety and nervousness.

2. Where will I park?

3. How does the exam site look? Will I have ample workspace? Will it be overly bright (sunglasses), or cold (sweater), or noisy (earplugs)? Would a cushion make the chair more comfortable?

4. Where is the drinking fountain? Lavatory facilities? Pay phone?

5. What about food? Most states do not allow food in the test room (exceptions for ADA). Should I take something along for energy in the exam? A light bag lunch during the break makes sense.

Items to Take to the Examination

Although you may not bring books to the exam, you should bring the following:

■ *Calculator*—Beginning with the April 2004 exam, NCEES has implemented a more stringent policy regarding permitted calculators. In brief, you may bring a battery-operated, silent, nonprinting, noncommunicating calculator. For more details, see the NCEES Web site (www.ncees.org), which includes the updated policy and a list of frequently asked questions about permitted calculators. You also need to determine whether your state permits pre-programmed calculators. Bring extra batteries for your calculator just in case, and many people feel that bringing a second calculator is also a very good idea.

■ *Clock*—You must have a time plan and a clock or wristwatch.

■ *Exam Assignment Paperwork*—Take along the letter assigning you to the exam at the specified location to prove that you are the registered person. Also bring something with your name and picture (driver's license or identification card).

■ *Items Suggested by Your Advance Visit*—If you visit the exam site, it will probably suggest an item or two that you need to add to your list.

■ *Clothes*—Plan to wear comfortable clothes. You probably will do better if you are slightly cool, so it is wise to wear layered clothing.

Special Medical Condition

If you have a medical situation that may require special accommodation, you need to notify the licensing board well in advance of exam day.

Examination Scoring

The questions are machine-scored by scanning. The answer sheets are checked for errors by computer. Marking two answers to a question, for example, will be detected and no credit will be given.

Geotechnical

Dr. Braja M. Das

PARTICLE SIZE DISTRIBUTION

The particle size distribution in a given soil is determined in the laboratory by sieve analysis and hydrometer analysis. For classification purposes, in coarse-grained soils the following two parameters can be obtained from a particle size distribution curve:

$$\text{Uniformity coefficient, } c_u = \frac{D_{60}}{D_{10}} \tag{1.1}$$

$$\text{Coefficient of gradation, } c_c = \frac{D_{30}^2}{D_{60} \times D_{10}} \tag{1.2}$$

where D_{10}, D_{30}, D_{60} = diameters through which, respectively, 10 percent, 30 percent, and 60 percent of the soil pass.

WEIGHT-VOLUME RELATIONSHIPS

Soils are three-phase systems containing soil solids, water, and air (Fig. 1.1).

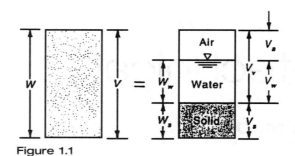

Figure 1.1

Referring to the figure,

$$W = W_s + W_w \tag{1.3}$$

$$V = V_s + V_v = V_s + V_w + V_a \tag{1.4}$$

where W = total weight of the soil specimen, W_s = weight of the solids, W_w = weight of water, V = total volume of the soil, V_s = volume of soil solids, V_v = volume of voids, V_w = volume of water, and V_a = volume of air.

The **volume relationships** can then be given as follows:

$$\text{Void ratio} = e = \frac{V_v}{V_s} \tag{1.5}$$

$$\text{Porosity} = n = \frac{V_v}{V} = \frac{e}{1+e} \tag{1.6}$$

$$\text{Degree of saturation} = S(\%) = \frac{V_w}{V_v} \times 100 = \frac{wG}{e} \times 100 \tag{1.7}$$

Similarly, the **weight relationships** are

$$\text{Water content (or moisture content)}(\%) = w = \frac{W_w}{W_s} \times 100 \tag{1.8}$$

$$\text{Moist unit weight} = \gamma = \frac{W}{V} = \frac{G\gamma_w(1+w)}{1+e} \tag{1.9}$$

$$\text{Dry unit weight} = \gamma_d = \frac{W_s}{V} = \frac{\gamma}{1+w} = \frac{G\gamma_w}{1+e} \tag{1.10}$$

$$\text{Unit weight of solids} = \gamma_s = \frac{W_s}{V_s} = G\gamma_w \tag{1.11}$$

In the preceding equations, γ_w = unit weight of water (62.4 lb/ft^3 or 9.81 kN/m^3) and G = specific gravity of soil solids, or

$$G = \frac{W_s}{V_s\gamma_w} \tag{1.12}$$

RELATIVE DENSITY

In granular soils, the degree of compaction is generally expressed by a nondimensional parameter called **relative density**, D_d, or

$$D_d = \frac{e_{max} - e}{e_{max} - e_{min}} \tag{1.13}$$

where e = actual void ratio in the field, e_{max} = void ratio in the loosest state, and e_{min} = void ratio in the densest state.

In terms of dry unit weight,

$$D_d = \frac{\dfrac{1}{\gamma_{min}} - \dfrac{1}{\gamma_d}}{\dfrac{1}{\gamma_{min}} - \dfrac{1}{\gamma_{max}}} \tag{1.14}$$

where γ_{min}, γ_{max} = minimum and maximum *dry* unit weights, respectively and γ_d = dry unit weight in the field.

CONSISTENCY OF CLAYEY SOILS

The moisture content in percent at which the cohesive soil will pass from a liquid state to a plastic state is called the **liquid limit**. Similarly, the moisture contents at which the soil changes from a plastic state to a semisolid state and from a semisolid state to a solid state are referred to as the **plastic limit** and the **shrinkage limit**, respectively. These limits are referred to as the **Atterberg limits** (see Fig. 1.2).

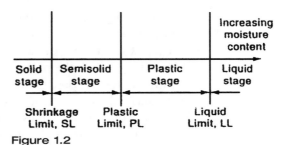

Figure 1.2

Liquid limit (LL) is the moisture content in percent at which the groove in the *Casagrande liquid limit device* closes for a distance of 0.5 in. after 25 blows.

Plastic limit (PL) is the moisture content in percent at which the soil, when rolled into a thread of 1/8 in. diameter, crumbles.

Plasticity index (PI) is defined as

$$PI = LL - PL \tag{1.15}$$

Shrinkage limit (SL) is the moisture content at which the volume of the soil mass no longer changes.

Shrinkage index (SI) is defined as

$$SI = PL - SL \tag{1.16}$$

PERMEABILITY

The rate of flow of water through a soil of gross cross-sectional area A can be given by the relationships shown in Fig. 1.3, called Darcy's law,

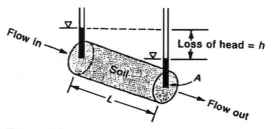

Figure 1.3

$$v = ki \tag{1.17}$$

where v = discharge velocity, k = coefficient of permeability, i = hydraulic gradient = h/L (see Fig. 1.3).

$$Q = vA = kiA \tag{1.18}$$

where Q = flow through soil in unit time and A = area of cross section of the soil at a right angle to the direction of flow. Or

$$k = \frac{Q}{iA} \tag{1.19}$$

FLOW NETS

In many cases, flow of water through soil varies in direction and in magnitude over the cross section. In those cases, calculation of rate of flow of water can be made by using a graph called a **flow net**. A flow net is a combination of a number of flow lines and equipotential lines. A flow line is one along which a water particle will travel from the upstream side to the downstream side. An equipotential line is one along which the potential head at all points is the same.

Figure 1.4 shows an example of a flow net in which water flows from the upstream to the downstream around a sheet pile. Note that in a flow net the flow lines and equipotential lines cross at *right angles*. Also, the flow elements constructed are *approximately square*. Referring to Fig. 1.4, the flow line in unit time

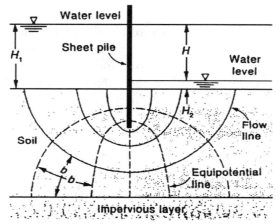

Figure 1.4

(Q) per unit length normal to the cross section shown is

$$Q = k \frac{N_f}{N_d} H \tag{1.20}$$

where N_f = number of flow channels, N_d = number of drops, and H = head difference between the upstream and downstream side. (Note that in Fig. 1.4, N_f = 4 and N_d = 6.)

CONSOLIDATION

Consolidation settlement is the result of volume change in saturated clayey soils due to the expulsion of water occupied in the void spaces. In soft clays the major portion of the settlement of a foundation may be due to consolidation. Based on the theory of consolidation, a soil may be divided into two major categories: (a) normally consolidated and (b) overconsolidated. For *normally consolidated* clay, the *present effective overburden pressure* is the maximum to which the soil has been subjected in the recent geologic past.

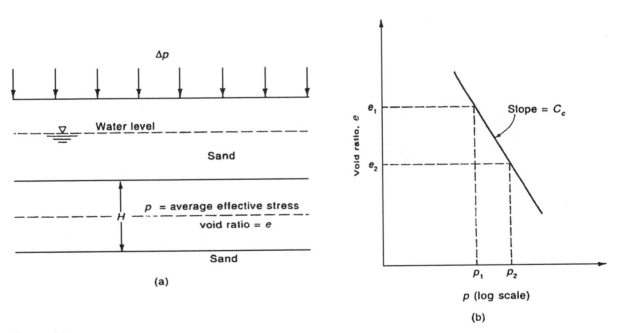

Figure 1.5

Figure 1.5(a) shows a normally consolidated clay deposit of thickness H. The consolidation settlement (ΔH) due to surcharge Δp can be determined as

$$\Delta H = \frac{C_c H}{1 + e_i} \log \left(\frac{p_i + \Delta p}{p_i} \right) \tag{1.21}$$

where e_i = initial void ratio, p_i = initial *average* effective pressure, and C_c = compression index.

The compression index can be determined from the laboratory as (see Fig. 1.5(b))

$$C_c = \frac{e_1 - e_2}{\log \left(\frac{p_2}{p_1} \right)} \tag{1.22}$$

SHEAR STRENGTH

The shear strength of soil (s), in general, is given by the **Mohr-Coulomb failure criteria**, or

$$s = c + \sigma' \tan \phi \tag{1.23}$$

where c = cohesion, σ' = effective normal stress, and ϕ = drained friction angle. For sands, $c = 0$, and the magnitude of ϕ varies with the relative density of compaction, size, and shape of the soil particles.

For normally consolidated clays, $c = 0$, so $s = \sigma' \tan \phi$. For overconsolidated clays, however, $c \neq 0$, so $s = c + \sigma' \tan \phi$. An important concept for the shear strength of cohesive soils is the so-called $\phi = 0$ concept. This is the condition in which drainage from the soil does not take place during loading. For such a case

$$s = c_u \tag{1.24}$$

where c_u is the undrained shear strength.

The **unconfined compression strength**, q_u, of a cohesive soil is

$$q_u = 2c_u \tag{1.25}$$

PROBLEMS

1.1 A moist soil specimen has a volume of 0.15 m³ and weighs 2.83 kN. The water content is 12 percent, and the specific gravity of soil solids is 2.69. Determine
   a. Moist unit weight, γ
   b. Dry unit weight, γ_d
   c. Void ratio, e
   d. Degree of saturation, S

1.2 For a soil deposit in the field, the dry unit weight is 14.9 kN/m³. From the laboratory, the following were determined: $G = 2.66$, $e_{max} = 0.89$, $e_{min} = 0.48$. Find the relative density in the field.

1.3 For a sandy soil, the maximum and minimum void ratios are 0.85 and 0.48, respectively. In the field, the relative density of compaction of the soil is 29.3 percent. Given $G = 2.65$, determine the moist unit weight of the soil at $w = 10\%$.

1.4 Refer to the flow net shown in Fig. 1.4. Given $k = 0.03$ cm/min, $H_1 = 10$ min, and $H_2 = 1.8$ m, determine the seepage loss per day per foot under the sheet pile construction.

1.5 The results of a sieve analysis of a granular soil are as follows:

U.S. Sieve No.	Sieve Opening (mm)	Percent Retained on Each Sieve
4	4.75	0
10	2.00	20
40	0.425	20
60	0.25	30
100	0.15	20
200	0.075	5

Determine the uniformity coefficient and coefficient of gradation of the soil.

1.6 For a normally consolidated clay of 3.2 m thickness, the following are given:

$$\text{Average effective pressure} = 98 \text{ kN/m}^2$$
$$\text{Initial void ratio} = 1.1$$
$$\text{Average increase of pressure in the clay layer} = 42 \text{ kN/m}^2$$
$$\text{Compression index} = 0.27$$

Estimate the consolidation settlement.

1.7   An oedometer test in a normally consolidated clay gave the following results.

Average Effective Pressure (kN/m^2)	Void Ratio
100	0.9
200	0.82

Calculate the compression index.

SOLUTIONS

1.1

a. $\gamma = \dfrac{W}{V} = \dfrac{2.83}{0.15} = 18.87 \text{ kN/m}^3$

b. $\gamma_d = \dfrac{\gamma}{1+w} = \dfrac{1887}{1+\left(\frac{12}{100}\right)} = 16.85 \text{ kN/m}^3$

c. $\gamma_d = \dfrac{G\gamma_w}{1+e}; \quad e = \dfrac{G\gamma_w}{\gamma_d} - 1 = \dfrac{(2.69)(9.81)}{16.85} - 1 = 0.566$

d. $S = \dfrac{wG}{e} \times 100 = \dfrac{(0.12)(2.69)}{0.566} \times 100 = 57.03\%$

1.2   In the field

$$\gamma_d = \dfrac{G\gamma_w}{1+e}; \quad e = \dfrac{G\gamma_w}{\gamma_d} - 1 = \dfrac{(2.66)(9.81)}{14.9} - 1 = 0.75$$

$$D_d = \dfrac{e_{max} - e}{e_{max} - e_{min}} = \dfrac{0.85 - 0.75}{0.89 - 0.48} = 34\%$$

1.3

$$D_d = 0.293 = \dfrac{e_{max} - e}{e_{max} - e_{min}} = \dfrac{0.85 - e}{0.85 - 0.45}; \quad e = 0.733$$

$$\gamma = \dfrac{G\gamma_w(1+w)}{1+e} = \dfrac{(2.65)(9.81)(1+0.1)}{1+0.733} = 16.5 \text{ kN/m}^3$$

1.4

$$Q = k\dfrac{N_f}{N_d}H = \dfrac{(0.03 \times 60 \times 24 \text{ cm/day})}{100}\left(\dfrac{4}{6}\right)(10-1.8) = 2.36 \text{ m}^3\text{/day/m}$$

1.5

Sieve Opening (mm)	Cumulative Percent Passing
4.75	100
2.0	80
0.425	60
0.25	30
0.15	10
0.075	5

So $D_{60} = 0.425$ mm; $D_{30} = 0.25$ mm; $D_{10} = 0.15$ mm

$$c_u = \frac{D_{60}}{D_{10}} = \frac{0.425}{0.15} = 2.83$$

$$c_c = \frac{D_{30}^2}{D_{60} \times D_{10}} = 0.98$$

1.6

$$\Delta H = \frac{C_c H}{1 + e_i} \log \left(\frac{p_i + \Delta p}{p_i} \right) = \frac{(0.27)(3)}{1 + 1.1} \log \left(\frac{98 + 42}{98} \right) = 0.0597 \text{ m} = 59.7 \text{ mm}$$

1.7

$$C_c = \frac{e_1 - e_2}{\log \left(\frac{p_2}{p_1} \right)} = \frac{0.9 - 0.82}{\log \left(\frac{200}{100} \right)} = 0.266$$

Structural Analysis

Dr. Michael Taylor

The outline notes that constitute this chapter are a summary of what can be found in all textbooks upon this subject. Since no one person's choice of summary can hope to match the needs of all students simultaneously, it is recommended that a suitable textbook be reread in conjunction with these notes and that personal notes be appended to the text in this chapter.

Determinate structural analysis, often termed **statics**, deals with structures that do not move. There are only two types of motion: translation and rotation. If the structure (and all parts of it) neither translates nor rotates, it is said to be in **static equilibrium**. (It is true, of course, that any material under stress will undergo some change of size and/or shape because of those stresses, but these movements are considered negligible in the present context.)

In statics, translation is caused by **forces**, and rotations are caused by **moments**. These are vector quantities. To define a vector requires three characteristics, usually (a) a line of action, (b) a direction, and (c) a magnitude. In some contexts, moments whose vectors are out of the plane are termed **torques** or **twists**. The student is expected to be familiar with simple vectors and their manipulation.

Actions is a general term that includes both forces and moments.

NEWTON'S LAWS

The study of statics is based upon two of Newton's three laws of motion. These are (1) a body will remain in its state of rest (relative to some chosen reference point) or of motion in a straight line, unless acted upon by a force, and (2) to every action there is an equal and opposite reaction. The third law (a body under the action of a single force will translate with acceleration along the line of the force and in the same direction as the force) is used in the study of *dynamics*.

Law 2 is often restated as "Forces (actions) can exist only in equal and opposite pairs." The combination of laws 1 and 2 thus requires that the net action on a body at rest be zero. This requirement, in turn, defines the two vector equations

$$\text{Net force} = 0$$
$$\text{Net moment} = 0$$

This is the central principle of statics.

More generally, determinate structural analysis can be defined as the process of calculating (for a body in static equilibrium)

(a) Any one, or all, of the actions acting upon the body

(b) The movement (deflection) of any point (in any direction) within the body

(c) The rotation of any line in the body

The process of structural analysis has just two steps. *First*, an idealized model of the structure (including the forces acting upon it) is agreed upon. In all but the most complex or large structures, it is adequate to use two-dimensional representations of real structures. When modeling the structure, it is also common practice to idealize the supports. Such idealizations as frictionless rollers, fully fixed supports, and frictionless pins are assumed to be familiar to the reader.

One other assumption is that structures are composed of rigid members. This does not mean that each member is infinitely rigid (i.e., does not change length or shape when loaded) but rather that these deformations are small relative to the original dimensions. Expressed another way, the deformations induced by loading do not change the geometry of the structure significantly. Luckily, this assumption is a good one for the vast majority of real engineering structures. The discerning reader may be about to object that many textbook examples (and reality) contain structures with ropes or wires, which are not stable if placed in compression. This objection is a good one but is countered by the observation (often implicit) that in all such cases the rope or wire must be in tension and hence cannot "buckle" or otherwise move.

Most engineers' drawings (and most examination problems) already include the simplifications just described.

The *second* step of the analysis process is the application of Newton's laws to the model.

The application of these equations of statics (often termed the equations of equilibrium) to a three-dimensional body involves just two vector equations (translation = 0 and rotation = 0). However, it is frequently more convenient to use the equivalent six independent scalar equations (three for translation and three for rotations). For two-dimensional models these reduce to three (two translations and one rotation). Thus, if a two-dimensional structural problem involves more than three unknowns, it is statically indeterminate and requires further knowledge for its solution. In what follows, the discussion is limited to two dimensions and stable static structures.

FREE BODY DIAGRAMS

The single most important concept in the *solution* of statics problems is that of the **free body diagram** (FBD). This concept is founded on the observation that if a (rigid) structure is in equilibrium, then every part of it must also be in equilibrium. This method has proven to minimize the likelihood of errors in solving structural problems.

In this approach an imaginary closed line is drawn through the structure. This line isolates a part of the structure, called the **free body**. The isolated part may be drawn by itself, but to maintain consistency with the original structure and loading, the engineer must examine the complete length of the closed line and insert all possible actions in their correct form and at their correct locations. Of course, some of these actions will be known and some unknown. All known actions must be included with their correct locations, magnitudes, and orientations. All unknown actions are given names, and their directions may be assumed (the mathematics will provide both the correct magnitude and the correct direction for each unknown).

It should be obvious that if the FBD is incorrect in *any way*, then the solution will be incorrect. A solution may be obtained, but if so, it is the solution to some *other* problem.

It is recommended that for each FBD used, a clear choice of axes be indicated and a positive direction also be chosen. These choices are, of course, arbitrary. It is then recommended that when applying the equations of equilibrium, all terms be written on one side of the equation and the result set equal to zero. The consistent use of this technique will reduce errors. By contrast, the practice, for example, of setting "all up forces equal to all down forces," which may seem attractive at first, is fraught with peril.

Facility in the analysis of structures is best obtained by solving multiple problems of as wide a variety as can be found. An excellent learning tool is to outline the steps of the solution without necessarily performing the mathematics. This permits a more rapid acquisition of the skills needed to understand how new problems are solved.

Note that in the following examples, units are not specified. The reader may select any units desired (e.g., metric versus "English" and small versus large).

In the following two problems the entire structure is used as the FBD.

Example 2.1

Find the reactions necessary to keep the structure shown in Exhibit 1 in equilibrium. Assume the axes, names, and directions shown in the figure.

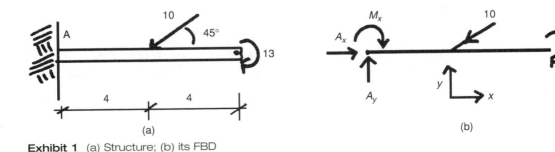

Exhibit 1 (a) Structure; (b) its FBD

Solution

Step 1. Sum forces in x direction:

$$A_x + (-).707 \times 10 = 0$$
$$A_x = +7.07 \text{ (assumed direction correct)}$$

Step 2. Sum forces in y direction:

$$A_y + (-).707 \times 10 = 0$$
$$A_y = +7.07 \text{ (assumed direction correct)}$$

Step 3. Sum moments about an axis through A (assume clockwise positive):

$$(+)M_x + (+)10 \times 4 \times .707 + (+)13 = 0$$
$$M_x = -41.3 \text{ (assumed direction was incorrect)}$$

This completes the solution requested.

Example **2.2**

Find the reactions necessary to keep the structure shown in Exhibit 2 in equilibrium.

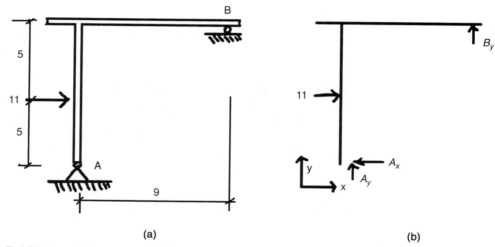

(a)   (b)

Exhibit 2   (a) Structure; (b) FBD of entire structure

Assume the names, axes, and directions shown.

Solution

Step 1. Summing forces in the positive x direction gives

$$+11 + (-)(A_x) = 0$$

Hence $(A_x) = +11$ (i.e., assumed direction is correct)

Step 2. Taking moments about an axis through A (with clockwise positive) gives

$$(-)B_y \times 9 + (+)11 \times 5 = 0$$

Hence

$$B_y = +\frac{55}{9} = +6.1$$

Step 3. Summing forces in the y direction gives

$$A_y + 6.1 = 0$$
$$A_y = -6.1 \text{ (i.e., assumed direction was incorrect)}$$

This completes the solution requested.

TRUSSES AND FRAMES

It should already be obvious to the reader that there is no single best way to solve structural analysis problems. It is true that a judicious choice of FBD and of the order in which equations are solved can shorten the calculation process, but all solution approaches will produce the correct answers if applied without error. However, it is convenient (but nothing more) to classify certain groups of structures, because they share characteristics and thus can be handled by a more "systematic" solution process. Two such groups are trusses and frames.

Trusses

Trusses must meet three criteria:

- All members (often termed **bars**) of the structure must be connected at only two points (these are called **joints**).

- All joints are frictionless pins.

- Loads are applied only at the joints.

The FBD of a bar of a truss can thus have only two forces acting upon it: one at each frictionless pin. Note that the bars need not be straight (although they almost always are). Application of the equations of statics shows that the line of action of both these forces must be directed along the line joining the two joints and that they must be equal in magnitude but oppositely directed. Hence a member of a truss is often described as a "two-force member." Only two possibilities then exist: Either the two forces are directed toward each other (in which case the bar is in compression) or they are directed away from each other (in which case the bar is in tension).

This feature of trusses simplifies their analysis, because if an FBD is drawn that cuts a member, it is known that only one action can exist at the cut, and it must be a force directed along the line that connects the joints (the sense and magnitude to be found as part of the solution).

There are two "classes" of analysis for trusses.

- **Method of Joints:** In this approach all FBDs are joints (with the exception that the entire body FBD is often used to find the reaction).

- **Method of Sections:** In this method the FBDs can include portions of the structure that are larger than a single joint. It should be noted that most designers use a combination of these methods when solving structures.

Example **2.3**

Find the bar force in member CB in Exhibit 3(a) using the method of joints. Use an FBD of the entire structure, shown in Exhibit 3(a), to obtain reactions. Assume the axes, names, and directions shown.

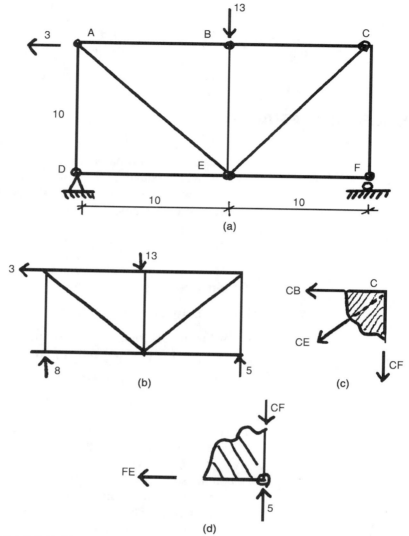

Exhibit 3 Figures for Example 2.3

Solution

Solving by the methods shown above, the reader should obtain

$$D_y = 8.0 \text{ up}$$
$$F_y = 5.0 \text{ up}$$
$$D_x = 3 \text{ to right}$$

This completes the reactions, as shown in Exhibit 3(b).

Note that if we wish to attempt to find CB using an FBD of joint C, there will be three unknowns, which are not solvable, see Exhibit 3(c). We must find one of the other bar forces, CE or CF, in order to obtain a solution for CB. One possibility is to solve joint F first. This will give us CF and then permit a solution at joint C.

With Exhibit 3(d) and the assumed directions, joint F is solved to give CF = 5 (with bar in compression).

Now we move back to joint C in Exhibit 3(c) but with CF known in magnitude and direction—hence the arrow for CF in Exhibit 3(c) must be reversed. Taking moments about E with clockwise assumed positive,

CB = 5 and CB is in compression

Example **2.4**

Find the bar force in member DH in Exhibit 4(a) using the method of sections.

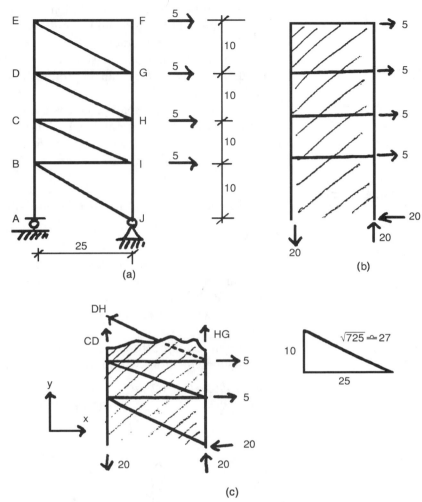

Exhibit 4 Figures for Example 2.4; (a) Structure; (b) FBD entire structure; (c) FBD with horizontal cut through DC and GH

Solve to get the reactions shown in Exhibit 4(b).

Solution

Note that a horizontal cut can be used to give the FBD shown in Exhibit 4(c). Sum forces in *x* direction:

$$+5 + 5 - 20 - DH \times \frac{25}{27} = 0$$

$$DH = +10.8 \text{ (assumed direction correct, bar DH in tension)}$$

Frames

Frames are multimember structures that do not meet the criteria for trusses (although individual members within a structure may be two-force members and thus help to simplify their solution). Thus, when an FBD cuts a member, there are (in two dimensions) three possible actions across the cut, and these must be included in the FBD.

Frames are clearly more complicated to solve than trusses, and there are no simple rules or guidelines for solution other than (a) the general rule that several FBDs will be needed for a solution and (b) the designer must experiment with several possible FBDs to find a solution.

It is an excellent practice (though very infrequently followed) to outline the steps of a solution (i.e., the sequence of FBDs to be used and the identification of the unknowns that can be found at each step) prior to performing any calculations.

When relatively small problems are solved, it is always a good idea to make a quick "survey" of the completed solution to check for gross errors. The equations of statics can often be assessed approximately without the need for pencil and paper.

Example 2.5

Determine the force in member AC in Exhibit 5(a).

Solution

Note that this structure is externally indeterminate; that is, the reactions cannot be found by the equations of statics alone. We note, however that bar AC is a two-force member, so we can take BCD and FBD and solve for the force AC. (Note that we cannot solve the second FBD completely, but we can get the bar force asked for.)

From Exhibit 5(b) and by taking moments about B (note that neither of the translation equations will produce a numerical result), with clockwise positive:

$$(+)17 \times 17 + (-).707 \times CA \times 11 = 0$$

$$CA = 37.2 \text{ (assumed direction correct, CA in tension)}$$

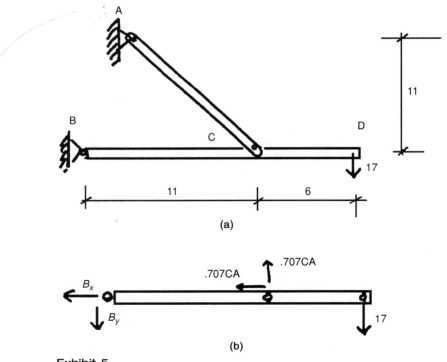

(a)

(b)

Exhibit 5

PROBLEMS

The reader is encouraged to attempt these problems alone prior to reviewing the sample solutions and to gain insight by comparing solutions where these differ.

In all cases the self-weight of the structure may be ignored unless instructions indicate otherwise.

Indicate the best answer from those offered (roundoff errors may lead to small discrepancies).

2.1 For the structure shown in Exhibit 2.1 the reactions are (in kN)

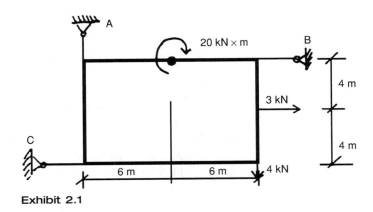

Exhibit 2.1

    a. A = 4 kN down, B = 10 kN to left, C = 11.5 kN to right
    b. A = 3 kN down, B = 10 kN to right, C = 10.0 kN to right
    c. A = 4 kN up, B = 10 kN to left, C = 7.0 kN to right
    d. A = 4 kN up, B = 8.5 kN to left, C = 11.5 kN to left

2.2 The truss shown in Exhibit 2.2 is used as a weighing device. If the gauge at A reads 9 kN, what is the weight of mass P?

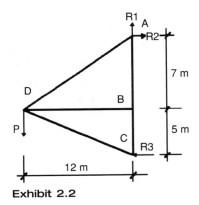

Exhibit 2.2

    a. 9 kN
    b. 12 kN
    c. 4 kN
    d. 8 kN

2.3   The force in member CB in Exhibit 2.3 is
   a.   4.65 kN tensile
   b.   4.65 kN compressive
   c.   47.5 kN tensile
   d.   8.00 kN compressive

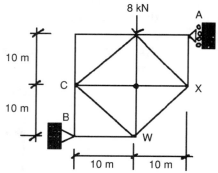

Exhibit 2.3

2.4   The force in member DE in Exhibit 2.4 is
   a.   11.2 kN tensile
   b.   11.3 kN compressive
   c.   Zero
   d.   None of the above

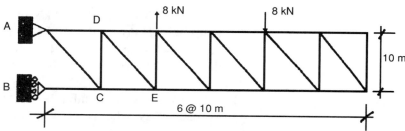

Exhibit 2.4

2.5   The forces in members FB and GH in Exhibit 2.5 are
   a.   FB = 2.5 kN tensile, GH = 11.66 kN tensile
   b.   FB = zero kN tensile, GH = 11.66 kN tensile
   c.   FB = zero kN tensile, GH = 11.66 kN compressive
   d.   FB = zero kN tensile, GH = 10.00 kN tensile

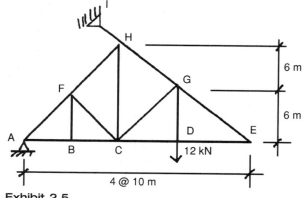

Exhibit 2.5

SOLUTIONS

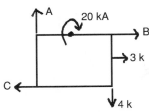

Exhibit 2.1a FBD whole
structure

2.1 c.

$$\Sigma V = 0 \text{ (positive upward)}$$
$$+A - 4 = 0$$
$$A = 4 \text{ upward}$$

$$\Sigma M_c = 0 \text{ (positive clockwise)}$$
$$+20 + B(8) + 3(4) + 4(12) = 0$$
$$B = -10$$
$$B = 10 \text{ leftward}$$

$$\Sigma H = 0 \text{ (positive rightward)}$$
$$-C(-)10 + 3 = 0$$
$$C = -7$$
$$C = 7 \text{ rightward}$$

2.2 a.

$$\Sigma M_c = 0 \text{ (positive clockwise)}$$
$$+9 \times 12 - P(12) = 0$$
$$P = 9k$$

2.3 d.

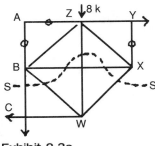

Exhibit 2.3a

Note: XY = 0 = ZA = AB.

$$\Sigma M_c = 0 \text{ (positive clockwise)}$$
$$+8(10) + R_Y(20) = 0$$
$$R_Y = 4 \leftarrow$$
$$\text{So } YZ = 4 \text{ comp.}$$
$$\Sigma H \text{ gives } C_H = 4 \text{ to the right}$$
$$\Sigma V \text{ gives } C_V = 8 \text{ upward}$$

So

$$CB = 8c$$
$$CW = 4c$$

2.4 c.

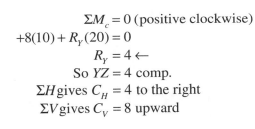

Exhibit 2.4a

FBD shown:

$$\Sigma V = 0$$
$$+8 - 8 + DE(0.7) = 0$$
$$DE = 0$$

2.5   b.

Joint B gives FB = 0

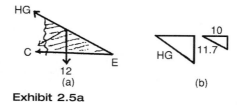

Exhibit 2.5a

FBD shown in Exhibit 2.5a:

$$\Sigma M_c = 0 \text{ (positive clockwise)}$$

$$+12 \times 10 - \left(\frac{10}{11.7}HG\right) \times 6$$

$$-\left(\frac{6}{11.7}HG\right) \times 10 = 0$$

$$120 - 5.13(HG) - 5.13(HG) = 0$$

$$HG = +\frac{120}{10.26} = 11.7$$

Hydraulics and Hydro Systems

Dr. Bruce E. Larock

OUTLINE

This chapter will review the two equations that are likely to be found in the FE examination. A more thorough presentation may be found in *Civil Engineering*: *License Review* or in selected references at the end of this chapter.

MANNING EQUATION

The Manning equation for the average velocity V in a steady open channel flow is

$$V = \frac{K}{n} R^{2/3} S^{1/2}$$

With SI units $K = 1.0$; with English units $K = 1.49$. The hydraulic radius is $R = A/P$, with $A =$ flow cross-sectional area and $P =$ wetted perimeter (i.e., the length of the interface along the fluid/solid boundary containing it). The slope S of the energy line is equal to the amount of head lost in a channel section of length L, or $S = h_L/L$. A short table of values for n, the Manning roughness factor, follows:

Channel Surface	Roughness Value, n
Concrete, finished	0.012
Concrete, gunite	0.019
Clay, vitrified sewer	0.014
Rubble masonry	0.025
Concrete, mortar	0.013
Concrete, troweled	0.013
Gravel, clean	0.025

Although the Manning equation is intended primarily for use in open channels, it is sometimes also used for pressurized flow in pipes.

Example **3.1**

A rectangular open channel lined with rubble masonry is 5 m wide and laid on a slope of 0.0004. If the depth of uniform flow is 3 m, compute the discharge in m^3/s.

Solution

From the table, the Manning roughness is approximately $n = 0.025$. The area and wetted perimeter are

$$A = (5)(3) = 15 \text{ m}^2$$
$$P = 5 + 2(3) = 11 \text{ m}$$
$$R = A/P$$

The Manning equation now gives

$$Q = AV = A \left(\frac{1}{n} \right) R^{2/3} S^{1/2}$$

$$= (15) \left(\frac{1}{0.025} \right) (15/11)^{2/3} (0.0004)^{1/2}$$

$$= 14.76 \text{ m}^3/s$$

Example **3.2**

A gunite concrete trapezoidal channel with 1:2 side slopes, is shown in Exhibit 1, conveys 60 m^3/s on a slope $S_o = 0.0005$. Compute the depth of uniform flow.

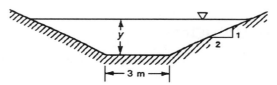

Exhibit 1

Solution

In SI units the Manning equation is

$$Q = (1/n) \, AR^{2/3} \, S_o^{2/3}$$

From the table the appropriate roughness coefficient is $n = 0.019$. Inserting the given information leads to

$$AR^{2/3} = 5.10 \qquad \text{(a)}$$

in which

$$A = 3y + 2y^2$$
$$P = 3 + 2\sqrt{5}y \qquad \text{and} \qquad R = A/P$$

Equation (a) must now must be solved by successive trial. It is convenient to use a table in doing so:

Trial	y (m)	A (m²)	P (m)	R	$R^{2/3}$	$AR^{2/3} = 5.10?$
1	1.5	9.00	9.71	0.93	0.95	8.55
2	1.0	5.00	7.47	0.67	0.77	3.85
3	1.15	6.10	8.14	0.75	0.83	5.06
4	1.16	6.17	8.19	0.75	0.83	5.12

The normal depth, that is, the depth of uniform flow, is $y = 1.16$ m.

HAZEN-WILLIAMS EQUATION

Many empirical formulas for pipe friction have been developed over the past century. These formulas are usually based on tests involving the flow of water under fully turbulent conditions and are not normally reliable for use with other fluids. One relatively widely used such formula is the Hazen-Williams equation, which is

$$V = 0.849 \, CR^{0.63} \, S^{0.54}$$

for SI units. For English units, replace 0.849 with 1.318. The Hazen-Williams coefficient C ranges from approximately 140 for very smooth and straight pipes to 120 for smooth masonry to 100 or less for old cast iron pipe. The other factors in the equation are defined as they were for the Manning equation.

Example 3.3

If 0.01 m³/s of water flows through a new 100 mm clean cast iron pipe ($C = 130$), determine the head loss in 1000 m of this pipe.

Solution

The discharge $Q = VA$ with

$$A = \frac{\pi}{4}(0.1)^2 = 0.00785 \, \text{m}^2$$

Hence $V = 0.01/0.00785 = 1.273$ m/s
   In this case the hydraulic radius is

$$R = \frac{\pi \frac{D^2}{4}}{\pi D} = \frac{D}{4} = \frac{0.1}{4} = 0.025 \text{ m}$$

The Hazen-Williams equation then yields

$$V = 1.273 = 0.849 \, (130)(0.025)^{0.63} \, S^{0.54}$$
$$S = 0.0191 = h_L/L = h_L/1000 \quad \text{and} \quad h_L = 19.1 \text{ m}$$

REFERENCES

Chow, V. T. *Open Channel Hydraulics*. McGraw-Hill, New York, 1959.

Henderson, F. M. *Open Channel Flow*. Macmillan, New York, 1966.

Street, R. L., Watters, G. Z., and Vennard, J. K. *Elementary Fluid Mechanics*, 7th Ed. Wiley, New York, 1996.

White, F. M. *Fluid Mechanics*, 3rd Ed. McGraw-Hill, New York, 1994.

Structural Steel and Reinforced Concrete Design

Dr. Alan Williams

ELASTIC DESIGN OF STEEL BEAMS

The allowable stress on flexural members depends on the shape of the section and the bracing used to prevent lateral instability.[1] Sections are classified as compact, noncompact, or slender in accordance with the criteria given in AISC Table B5.1. Most rolled W shapes qualify as compact sections. The exceptions are indicated in the AISC Tables of Properties by denoting the value of the yield stress F_y' and F_y'' at which a particular shape becomes noncompact.

Bending Stresses

The allowable bending stress for compact symmetrical shapes is given by AISC Equation (F1-1) as

$$F_b = 0.66F_y$$

when the maximum unbraced length of the compression flange does not exceed the smaller of

$$L_c = \frac{76b_f}{\sqrt{F_y}} \quad \text{or} \quad \frac{20,000}{(dF_y/A_f)}$$

where b_f = flange width and A_f = compression flange area.

When the unbraced length of a compact shape exceeds L_c but is less than the greater of

$$L_u = r_T\sqrt{102,000C_b/F_y} \quad \text{or} \quad \frac{20,000C_b}{(dF_y/A_f)}$$

the allowable bending stress is given by AISC Section F1.3 as

$$F_b = 0.60\ F_y$$

where r_T = radius of gyration of the compression flange plus one-third of the web area. The bending coefficient C_b is defined in Section F1.3 and Table 6 as

$$C_b = 1.75 + 1.05\ M_1/M_2 + 0.3(M_1/M_2)^2 \quad \text{but not more than 2.3}$$

The value of C_b conservatively may be taken as unity. Examples of the derivation of C_b are illustrated in Fig. 4.1. Values of L_c and L_u are tabulated in the AISC Beam Tables, assuming a value for C_b of unity.

In general, the allowable bending stress for noncompact symmetrical shapes is given by AISC Equation (F1-5) as

$$F_b = 0.60F_y$$

when the maximum unbraced length of the compression flange does not exceed L_c.

The AISC Selection Tables and Beam Tables tabulate the allowable beam resisting moments and uniformly distributed loads for W, M, S, C and MC shapes that are braced at the appropriate values of L_c or L_u. Adequate bracing is assumed to be provided by a restraint with a capacity of 1 percent of the force in the compression flange, in accordance with AISC Section G4. Note that metric steel tables are not available, so it is not possible to metricate the steel section.

When the unbraced length of the compression flange exceeds L_u, the allowable bending stress for both compact and noncompact shapes is given by the larger value from AISC Equations (F1-6), (F1-7), and (F1-8), but it may not exceed $0.60F_y$. When the unbraced length is less than

$$l = r_T\sqrt{510,000C_b/F_y} = L_1$$

the applicable equation is (F1-6), and the allowable stress is

$$F_b = F_y\left(0.667 - F_y l^2/1,530,000 r_T^2 C_b\right)$$

When the unbraced length equals or exceeds L_1, the applicable AISC equation is (F1-7), and the allowable stress is

$$F_b = \frac{170,000C_b r_T^2}{l^2}$$

| Configuration | Moment Diagram | M_1/M_2 | C_b |

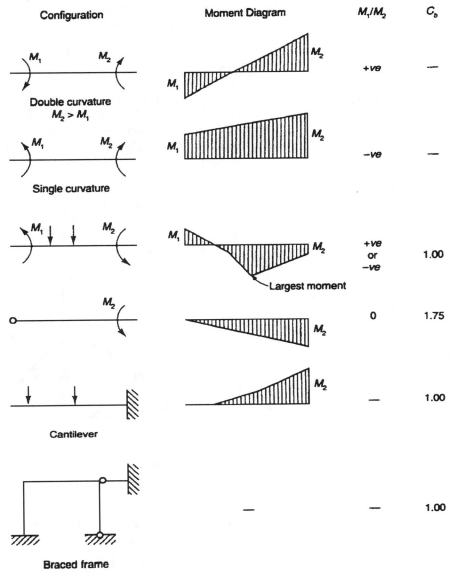

Figure 4.1 Derivation of C_b

Equation (F1-8) is independent of l/r_τ and gives an allowable stress of

$$F_b = \frac{12,000C_bA_f}{ld}$$

These expressions may be solved by calculator, or the allowable beam resisting moment may be obtained from AISC Moment Charts, which are based on a value of unity for C_b and are conservative for larger values of C_b.

Compact sections, solid rectangular sections bent about their weak axes, and solid round or square bars have an allowable stress given by AISC Equation (F2-1) of

$$F_b = 0.75F_y$$

Noncompact sections bent about their weak axes have an allowable stress given by AISC Equation (F2-2) of

$$F_b = 0.60F_y$$

| Example **4.1** |

The W18 × 60 grade A36 beam shown in Exhibit 1 is laterally supported throughout its length. Determine whether the beam is adequate to support the applied loads indicated. The relevant properties of the beam are $S_x = 108$ in^3, $F_y = 36$ kips/in^2, and allowable bending stress $F_b = 0.66F_y$.

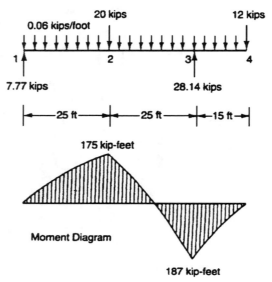

Exhibit 1

Solution

The bending moments acting on the beam from the applied loads and beam self-weight are shown in Exhibit 1.

$$M_x = \text{maximum moment} = 187 \text{ kip-ft}$$

$$f_b = \text{maximum bending stress} = M_x/S_x = 187 \times 12/108 = 20.78 \text{ kips/in}^2$$

The allowable stress is

$$F_b = 0.66F_y = 0.66 \times 36 = 23.76 \text{ kips/in}^2$$

Hence, the W18 × 60 is adequate.

Shear Stress

The allowable shear stress, based on the overall beam depth, is given by AISC Equation (F4-1) as

$$F_v = 0.40F_y$$

provided

$$h/t_w \le 380/\sqrt{F_y}$$

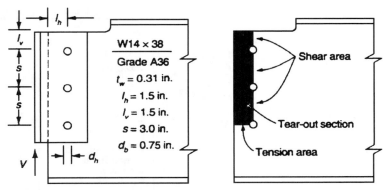

Figure 4.2 Block shear in a coped beam

and the actual shear stress is determined by

$$f_v = V/dt_w$$

where h = clear distance between the flanges, t_w = web thickness, d = overall depth of beam, and V = applied shear force.

When the end of the beam is coped, failure occurs by block shear, or web tear-out, which is a combination of shear along a vertical plane and tension along a horizontal plane. The resistance to block shear is given by

$$V_B = A_v F_v + A_t F_t$$

where A_v = net shear area, A_t = net tension area, F_v = allowable shear stress = $0.30F_u$ from AISC Equation (J4-1), and F_t = allowable tensile stress = $0.50F_u$ from AISC Equation (J4-2).

From Fig. 4.2, where $d_h = d_b + 0.0625$ in.

$$A_v = t_w(l_v + 2s - 2.5d_h) = 0.31[1.5 + 6 - 2.5(0.75 + 0.0625)] = 1.70 \text{ in}^2$$
$$A_t = t_w(l_h - 0.5d_h) = 0.31(1.5 - 0.5 \times 0.8125) = 0.34 \text{ in}^2$$

For grade A36 steel, F_u = 58 kips per square inch.

The resistance to block shear is then

$$V_B = A_v F_v + A_t F_t = 1.7 \times 0.30 \times 58 + 0.34 \times 0.50 \times 58 = 39.44 \text{ kips}$$

COMPRESSION MEMBERS

The allowable stress in an axially loaded compression member is dependent on the slenderness ratio, which is defined in AISC Section E2 as Kl/r, where r = the governing radius of gyration, Kl = effective length of the member, K = effective-length factor, and l = unbraced length of the member.

The value of the effective-length factor depends on the restraint conditions at each end of the column. AISC Table C-C2.1 specifies effective-length factors for well-defined standard conditions of restraint, and these are illustrated in Fig. 4.3. Values are indicated for ideal and practical end conditions, allowing for the fact that full fixity may not be realized. These values may be used only in simple cases when the tabulated end conditions are approached in practice.

End Restraints	Ideal K	Practical K	Shape
Fixed at both ends	0.5	0.65	
Fixed at one end, pinned at the other end	0.7	0.8	
Pinned at both ends	1.0	1.0	
Fixed at one end with the other end fixed in direction but not held in position	1.0	1.2	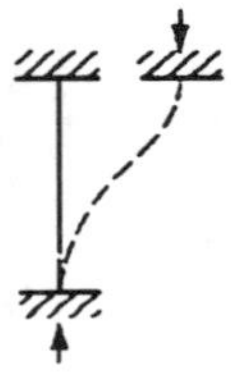
Pinned at one end with the other end fixed in direction but not held in position	2.0	2.0	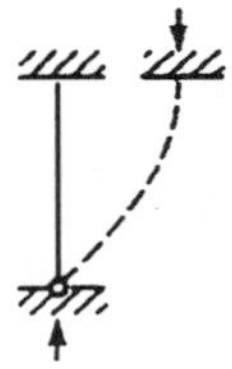
Fixed at one end with the other end free	2.0	2.1	

Figure 4.3 Effective-length factors

For compression members in a plane truss, AISC Section C-C2 specifies an effective-length factor of 1.0. For load-bearing web stiffeners on a girder, AISC Section K1.6 specifies an effective-length factor of 0.75. For columns in a rigid frame that is adequately braced, AISC Section C2.1 specifies a conservative value for the effective-length factor of 1.0.

The failure of a short, stocky column occurs at the squash load, when the strut yields in direct compression. The allowable stress in the strut is obtained from AISC Equation (E2-1) as

$$F_a = 0.6F_y$$

As the slenderness ratio of the column is increased, the failure load reduces. The Euler elastic critical load is assumed to govern when the column stress equals half the yield stress. The critical slenderness ratio, corresponding to this limit, is given by AISC Equation (C-E2-1) as

$$C_c = \sqrt{2\pi^2 \frac{E}{F_y}}$$

When the slenderness ratio exceeds this value, the allowable stress is

$$F_a = 12\pi^2 E/23(Kl/r)^2$$

When the slenderness ratio does not exceed this value, the allowable stress may be obtained from the expression

$$F_a = \left[1 - (Kl/r)^2/2C_c^2\right]F_y/\left[5/3 + 3(Kl/r)/8C_c - (Kl/r)^3/8C_c^3\right]$$

In accordance with AISC Section B7, the slenderness ratio should preferably not exceed 200.

Example **4.2**

The W14 × 120 Grade A36 column shown in Exhibit 2 is fixed at the base and unbraced about the x-axis. About the y-axis the column is held in position, at the top and at mid-height, but it is not fixed in direction. Determine whether the column is adequate to support an axial load of 500 kips.

The relevant properties of the W 14 × 120 are $A = 35.3$ in^2, $r_x = 6.24$ in., $r_y = 3.74$ in., $E_s = 29,000$ ksi, $F_y = 36$ ksi, $K_y = 0.8$, and $K_x = 2.1$.

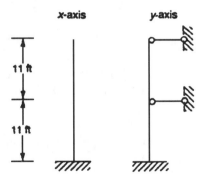

Exhibit 2   Column in Example 4.2

Solution

From Fig. 4.3, the slenderness ratio about the y-axis is given by

$$Kl/r_y = 0.8 \times 11 \times 12/3.74 = 28.2$$

The slenderness ratio about the x-axis is given by

$$Kl/r_x = 2.1 \times 22 \times 12/6.24 = 88.8, \text{ which governs}$$

$$C_c = \sqrt{2\pi^2 \times 29,000/36} = 126.1 > kl/r$$

The allowable axial stress is

$$F_a = [1 - 88.8^2/(2 \times 126.1^2)]36/[5/3 + 3 \times 88.8/(8 \times 126.1) - 88.8^3/(8 \times 126.1^3)] = 14.34 \text{ ksi}$$

The axial stress from the imposed loading is

$$f_a = P/A = 500/35.3 = 14.16 \text{ kips/in}^2 < F_a. \text{ The column is adequate.}$$

TENSILE STRESS

In determining the capacity of a connection in direct tension, as shown in Fig. 4.4, allowance must be made for the effective areas of the members and their method of attachment. To prevent excessive elongation of the members, which may lead to instability of the whole structure, AISC Section D1 limits the maximum tensile force on the connection to

$$P_t = 0.6F_y A_g$$

where A_g = gross area of the member = bt.

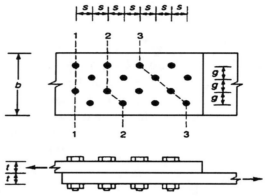

Figure 4.4 Net area of tension member

To prevent fracture of the member at the section of weakest effective net area, AISC Section D1 limits the maximum tensile force on the connection to

$$P_t = 0.5F_u A_e$$

where A_e = effective net area. The effective net area is defined in AISC Section B2 as

$$A_e = t(b - 2d_h) \qquad \text{for the section 1-1 in Fig. 4.4}$$

where d_h = specified diameter of hole = d_b + 1/8 in., where d_b = bolt diameter;

$$A_e = t(b - 3d_h + s^2/4g) \qquad \text{for section 2-2 in Fig. 4.4}$$

where s = longitudinal pitch and g = transverse gage;

$$A_e = t(b - 4d_h + 3s^2/4g) \qquad \text{for section 3-3 in Fig. 4.4}$$

To account for the effects of eccentricity and shear lag in rolled structural shapes connected through only part of their cross-sectional elements, the effective net area is given by AISC Equation (B3-1) as

$$A_e = UA_n$$

where A_n = net area of the member, U = 0.90 for I-sections with $b_f \geq 2d/3$ and with not less than three bolts in the direction of stress, U = 0.85 for all other shapes with not less than three bolts in the direction of stress, and $U = 0.75$ for all shapes with only two bolts in the direction of stress.

In addition, AISC Section B3 specifies that $A_e \leq 0.85A_g$.

STRENGTH DESIGN PRINCIPLES FOR REINFORCED CONCRETE MEMBERS

The basic requirement of designing for strength is to ensure that the design strength of a member is not less than the required ultimate strength. The latter consists of the service-level loads multiplied by appropriate load factors, and this is defined in ACI Equations[2] (9-1), (9-2), and (9-3) as

$$U = 1.4D + 1.7L$$
$$U = 0.75(1.4D + 1.7L + 1.7W)$$
$$U = 0.9D + 1.3W$$

where D = dead load, L = live load, and W = wind load.

The design strength of a member consists of the theoretical ultimate strength of the member—the **nominal strength**—multiplied by the appropriate strength reduction factor, ϕ. Thus

$$\phi \, (\text{nominal strength}) \geq U$$

ACI Section 9.3 defines the reduction factor as $\phi = 0.90$ for flexure, $\phi = 0.85$ for shear and torsion, $\phi = 0.75$ for compression members with spiral reinforcement, $\phi = 0.70$ for compression members with lateral ties, and $\phi = 0.70$ for bearing on concrete.

Flexure of Reinforced Concrete Beams

The nominal strength of a rectangular beam, with tension reinforcement only, is derived from the assumed ultimate conditions shown in Fig. 4.5. ACI Section 10.2.7.1 specifies an equivalent rectangular stress block in the concrete of $0.85f_c'$, with a depth of

$$a = A_s f_y / 0.85 f_c' b = \beta_1 c$$

where c = depth to neutral axis and β_1 = compression zone factor, given in ACI Section 10.2.7.3.

From Fig. 4.5, the nominal strength of the member is derived as

$$M_n = A_s f_y d (1 - 0.59 \rho f_y / f_c')$$

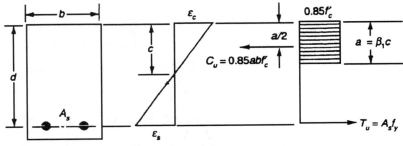

Figure 4.5 Member with tension reinforcement only

where $\rho = A_s/b_d$ = reinforcement ratio, $\phi M_n = 0.9M_n$ = design strength, and $\phi M_n \geq M_u$ = applied factored moment. This expression may also be rearranged to give the reinforcement ratio required to provide a given factored moment, M_u, as

$$\rho = 0.85f_c'\left[1 - \sqrt{1 - K/0.383f_c'}\right]/f_y$$

where $K = M_u/bd^2$.

These expressions may be readily applied using standard calculator programs[5] and tables.[3,4] For a balanced strain condition, the maximum strain in the concrete—and in the tension reinforcement—must simultaneously reach the values specified in ACI Section 10.3.2 as

$$\varepsilon_c = 0.003 = \text{concrete strain}$$
$$\varepsilon_s = f_y/E_s = \text{steel strain}$$

The balanced reinforcement ratio is given as $\rho_b = 0.85 \times 600\, \beta_1 f_c'/f_y(600 + f_y)$.

In accordance with ACI Section 10.3.3, the maximum allowable reinforcement ratio to ensure a ductile flexural failure with adequate warning of impending failure is

$$\rho_{max} = 0.75\rho_b$$

The maximum allowable reinforcement area is $A_{max} = bd\rho_{max}$.

The maximum allowable design strength of a singly reinforced member is

$$M_{max} = \phi A_{max}f_y\, d(1 - 0.59\rho_{max}f_y/f_c')$$

In accordance with ACI Sections 10.5.1. and 10.5.2, the minimum allowable reinforcement ratio is given by

$$\rho_{min} = 1.4/f_y$$

with the exception that the minimum reinforcement provided need not exceed one-third more than required by analysis.

Example 4.3

A reinforced concrete beam, with an overall depth of 400 mm, an effective depth of 350 mm, and a width of 300 mm, is reinforced with Grade 400M bars and has a concrete cylinder strength of 21 MPa. Determine the area of tension reinforcement required for the beam to support a superimposed live load of 15 kN/m over an effective span of 6 m.

Solution

The weight of the beam is $w_D = 0.4 \times 0.3 \times 23.5 = 2.82$ kN/m.

The dead load moment is given by $M_D = w_D P^2/8 = 2.82 \times 6^2/8 = 12.7$ kNm.

The live load moment is given by $M_L = w_L P^2/8 = 15 \times 6^2/8 = 67.5$ kNm.

The factored moment is $M_u = 1.4M_D + 1.7M_L = 1.4 \times 12.7 + 1.7 \times 67.5 = 132.5$ kNm.

The moment factor is $K = M_u/bd^2 = 132.5 \times 10^6/(300 \times 350^2) = 3.605$ MPa.

The reinforcement ratio required to provide a given factored moment M_u is

$$\rho = 0.85\, f_c'\, [1 - (1 - K/0.383\, f_c')^{0.5}]/f_y$$
$$= 0.85 \times 21\{1 - [1 - 3.605/(0.383 \times 21)]^{0.5}\}/400$$
$$= 0.011$$

The minimum allowable ratio is

$$\rho_{min} = 1.4/f_y = 1.4/400 = 0.0035 < \rho \qquad \text{Satisfactory}$$

The maximum allowable reinforcement ratio is

$$\rho_b = 0.75 \times 0.85 \times 600\beta 1\, f_c'/f_y(600 + f_y)$$
$$= 0.75 \times 0.85 \times 600 \times 0.85 \times 21/400(600 + 400)$$
$$= 0.017 > \rho \qquad \text{Satisfactory}$$

Hence the section size is adequate.

The reinforcement area required is $A_s = \rho_b d = 0.011 \times 300 \times 350 = 1155\ \text{mm}^2$

When the applied factored moment exceeds the maximum design strength of a singly reinforced member that has the maximum allowable reinforcement ratio, compression reinforcement and additional tensile reinforcement must be provided, as shown in Fig. 4.6. The difference between the applied factored moment and the maximum design moment strength of a singly reinforced section is $M_r = M_u - M_{max}$ = residual moment.

The additional area of tensile reinforcement required is

$$A_T = M_r/\phi f_y(d - d') = A_s' f_s'/f_y$$

The depth of the stress block is

$$a = f_y A_{max}/0.85\, f_c' b$$

The depth of the neutral axis is

$$c = a/\beta_1$$

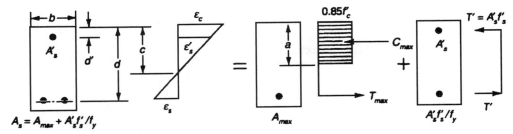

$A_s = A_{max} + A_s' f_s'/f_y$

Figure 4.6 Member with compression reinforcement

The stress in the compression reinforcement is given by[5]

$$f_s' = 600(1 - d'/c) \leq f_y$$

The required area of compression reinforcement is given by

$$A_s' = M_r/\phi\, f_s'(d - d')$$

The total required area of tension reinforcement is

$$A_s = A_{max} + A'_s f'_s / f_y$$

The maximum allowable reinforcement ratio is given by Section 10.3.3 as

$$\rho'_{max} = 0.75\rho_b + A'_s f'_s / bd f_y$$

In order to analyze a given member with compression reinforcement, an initial estimate of the neutral axis depth is required. The total compressive force in the concrete and compression reinforcement is then compared with the tensile force in the tension reinforcement. The initial estimate of the neutral axis depth is then adjusted until these two values are equal.

The conditions at ultimate load in a flanged member, when the depth of the equivalent rectangular stress block exceeds the flange thickness, are shown in Fig. 4.7. The area of reinforcement required to balance the compressive force in the flange is given by

$$A_{sf} = C_f / f_y = 0.85 f'_c h_f (b - b_w) / f_y$$

The corresponding design moment strength is

$$M_f = \phi A_{sf} f_y (d - h_f / 2)$$

The residual moment is $M_r = M_u - M_f$. The required reinforcement ratio to provide the residual moment is

$$\rho_w = 0.85 f'_c \left[1 - \sqrt{1 - 2K_w / 0.9 \times 0.85 f'_c} \right] / f_y$$

where $K_w = M_r / b_w d^2$.

The corresponding reinforcement area is

$$A_{sw} = b_u d \rho_w$$

The total reinforcement area required is

$$A_s = A_{sf} + A_{sw}$$

The total reinforcement ratio is

$$\rho = A_s / bd$$

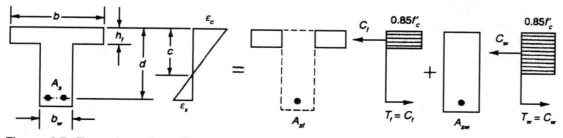

Figure 4.7 Flanged member with tension reinforcement

The maximum allowable reinforcement ratio is given by ACI Section 10.3.3 as

$$\rho'_{max} = 0.75 b_w (\rho_b + \rho_f)/b$$

where $\rho_b = 0.85 \times 600 \beta_1 f_c'/f_y(600+f_y)$, and $\rho_f = A_{sf}/b_w d$.

| Example **4.4** | The reinforced concrete beam shown in Exhibit 3 is reinforced with Grade 400M bars at the positions indicated and has a concrete cylinder strength of 21 MPa. The beam carries a superimposed load of 30 kN/m run over an effective span of 6 m. Determine the areas of tension and compression steel required. |

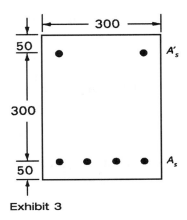

Exhibit 3

Solution

The weight of the beam is $w_D = 0.4 \times 0.3 \times 23.5 = 2.82$ kN/m. The dead load moment is given by $M_D = w_D l^2/8 = 2.82 \times 6^2/8 = 12.7$ kNm. The live load moment is given by $M_L = w_L l^2/8 = 30 \times 6^2/8 = 135.0$ kNm. The factored moment at midspan is obtained from ACI Equation (9-1) as

$$M_u = 1.4 M_D + 1.7 M_L = 1.4 \times 12.7 + 1.7 \times 135.0 = 247 \text{ kNm}$$

The compression zone factor is given by ACI Section 10.2.7. as $\beta_1 = 0.85$. The maximum allowable reinforcement ratio for a singly reinforced beam is

$$\rho_{max} = 0.75 \times 0.85 \times 600 \times \beta_1 f_c'/f_y(600+f_y)$$
$$= 0.75 \times 0.85 \times 600 \times 0.85 \times 21/400(600+400)$$
$$= 0.017$$

The maximum reinforcement area for a singly reinforced beam is

$$A_{max} = bd\,\rho_{max} = 300 \times 400 \times 0.017 = 2040 \text{ mm}^2$$

The maximum design moment of a singly reinforced section is

$$M_{max} = \phi A_{max} f_y d(1 - 0.59\,\rho_{max} f_y/f_c')$$
$$= 0.9 \times 2040 \times 400 \times 135(1 - 0.59 \times 0.017 \times 400/21)10^6 = 208 \text{ kNm}$$

The residual moment is given by

$$M_r = M_u - M_{max} = 39 \text{ kNm}$$

The depth of the stress block is

$$a = f_y A_{max}/0.85 f'_c b$$
$$= 400 \times 2040/(0.85 \times 21 \times 300)$$
$$= 152 \text{ mm}$$

The neutral axis depth is

$$c = a/\beta_l = 152/0.85 = 179 \text{ mm}$$

The stress in the compression reinforcement is

$$f'_s = 600(1 - d'/c) = 600 - (1 - 50/179)$$
$$= 400 \text{ MPa maximum}$$

The required area of compression reinforcement is

$$A'_s = M_r/\phi_s f'_s(d - d') = 106 \times 39/0.9 \times 400 \times 300$$
$$= 361 \text{ mm}^2$$

The total required area of tension reinforcement is

$$A_s = A_{max} + A'_s f'_s/f_y = 2040 + 361$$
$$= 2401 \text{ mm}^2$$

Deflection Requirements

Allowable deflections are given in ACI Table 9.5(b). For reinforced concrete members not supporting deflection-sensitive construction, the allowable deflection may be deemed satisfied if the minimum thickness requirements of ACI Tables 9.5(a) and 9.5(c) are adopted.

For other conditions, the short-term deflection may be computed from the effective moment of inertia given by ACI Equation (9-7) and indicated in Fig. 4.8 as

$$I_e = (M_{cr}/M_a)^3 I_g + [1 - (M_{cr}/M_a)^3]I_{cr}$$

where I_g = moment of inertia of gross concrete section, neglecting reinforcement = $bh^3/12$; I_{cr} = moment of inertia of the cracked transformed section = $b(kd)^3/3 + nA_s(d - kd)^2$; k = neutral axis depth ratio at service load for a singly reinforced section = $[2\rho n + (\rho n)^2]^{0.5} - \rho n$; $n = E_s/E_c$ = modular ratio; $\rho = A_s/bd$ = reinforcement ratio; M_a = maximum moment in the member; M_{cr} = cracking moment of the section $2f_r I_g/h$; and $f_r = 0.62\sqrt{f'_c}$ = modulus of rupture of normal-weight concrete.

When the member is subjected to long-term loads, creep and shrinkage produce additional deflection, which can be determined by multiplying the short-term deflection by the factor

$$\lambda = \xi/(1 + 50\rho')$$

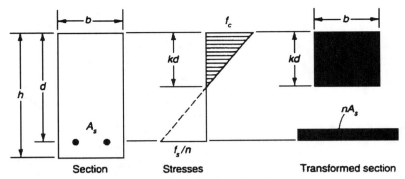

Figure 4.8 Reinforced member at service load

where ξ = time-dependent factor for sustained loads given in ACI Section 9.5.2.5, and $\rho' = A_s'/b_w d$ = reinforcement ratio for compression reinforcement.

Shear in Reinforced Concrete Members

The factored shear force acting on a member, in accordance with ACI Equations (11-1) and (11-2), is resisted by the combined design shear strength of the concrete and shear reinforcement. Thus, the factored applied shear is given by

$$V_u = \phi V_c + \phi V_s$$

where $\phi = 0.85$ = strength reduction factor for shear given by ACI Section 9.3.2.3; $V_c = 0.166\sqrt{f_c'}b_w d$ = nominal shear strength of normal weight concrete from ACI (11-3); $V_s = A_v f_y d/s$ = nominal shear strength of shear reinforcement from ACI Equation (11-17); A_v = area of shear reinforcement; and s = spacing of shear reinforcement, specified in ACI Section 11.5.4 as

$$s \leq d/2 \text{ or } 610 \text{ mm when } V_s \leq 0.33\sqrt{f_c'}b_w d,$$
$$s \leq d/4 \text{ or } 305 \text{ mm when } 0.33\sqrt{f_c'}b_w d < V_s \leq 0.66\sqrt{f_c'}b_w d$$

The dimensions of the section or the strength of the concrete must be increased, in accordance with ACI Section 11.5.6.8, to ensure that $V_s > 0.66\sqrt{f_c'}\,b_w d$

As specified in ACI Section 11.5.5, a minimum area of shear reinforcement is required when

$$V_u > \phi V_c/2 = V_{u(min)}$$

and the minimum area of shear reinforcement is given by ACI Equation (11-14) as

$$A_{v(min)} = 0.34 b_w s/f_y$$

When the support reaction produces a compressive stress in the member, ACI Section 11.1.3.1 specifies that the critical factored shear force is the force that is acting at a distance equal to the effective depth from the face of the support. Figure 4.9 summarizes the shear provisions of the Code, and Fig. 4.10 illustrates the design principles involved.

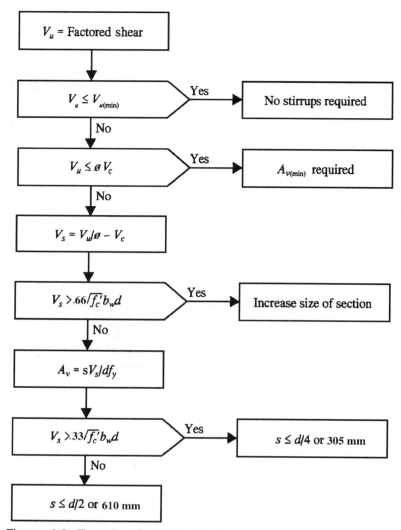

Figure 4.9   Flow chart for design for shear

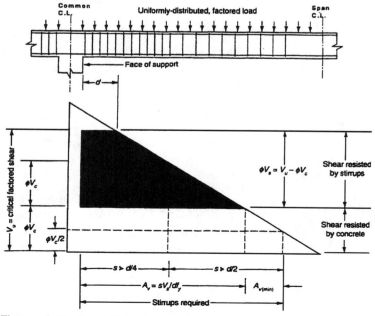

Figure 4.10   Shear in a reinforced concrete beam

Example 4.5

A reinforced concrete beam, which has an effective depth of 600 mm and a width of 500 mm, in reinforced with Grade 400M bars and has a compressive strength of 21 MPa. The factored shear force, at four locations on the beam, is (a) 800 kN, (b) 400 kN, (c) 100 kN, and (d) 500 kN. Determine the required spacing, at each location, for No. M10 stirrups with two or four vertical legs as suitable.

Solution

The design shear strength provided by the concrete is

$$\phi V_c = 0.166 \; \phi b_w d (f_c')0.5$$
$$= 0.166 \times 0.85 \times 500 \times 600(21)^{0.5}/1000$$
$$= 194 \text{ kN}$$

(a)

The design shear strength required from the shear reinforcement is

$$\phi V_s = V_u - \phi V_c = 800 - 194 = 606 \text{ kN}$$

Since $\phi V_s < 4 \times \phi V_c$ the concrete section is adequate.
Since $\phi V_s > 2 \times \phi V_c$ the maximum stirrup spacing is given by

$$s = d/4 = 600/4 = 150 \text{ mm}$$

The area of shear reinforcement required is

$$A_v/s = \phi V_s/\phi d f_y = 606 \times 10^6/(0.85 \times 600 \times 400)$$
$$= 2970 \text{ mm}^2/\text{m}$$

The required spacing of stirrups with four No. M10 vertical legs is

$$s = 4 \times 100/2.97 = 135 \text{ mm} < 150 \text{ mm} \qquad \text{Satisfactory}$$

(b)

The design shear strength required from the shear reinforcement is

$$\phi V_s = V_u - \phi V_c = 400 - 194 = 206 \text{ kN}$$

Since $\phi V_s < 2 \times \phi V_c$, the maximum stirrup spacing is given by

$$s = d/2 = 600/2 = 300 \text{ mm}$$

Reinforced Concrete Columns

Reinforced concrete columns may be classified as either short columns or long columns. For a short column, the axial load carrying capacity may be illustrated by reference to the column shown in Fig. 4.11.

The theoretical design axial load strength at zero eccentricity is given by ACI Section R10.3.5 as

$$\phi P_0 = \phi[0.85 f_c'(A_g - A_{st}) + A_{st} f_y]$$

where

ϕ = strength reduction factor specified in Section 9.3
 = 0.75 for compression members with spiral reinforcement and
 = 0.70 for compression members with lateral ties

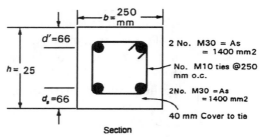

Figure 4.11   Compression in a short tied column

f'_c = concrete cylinder strength = 21 MPa
f_y = reinforcement yield strength = 400 MPa
A_g = gross area of the section = 62,500 mm^2
A_{st} = reinforcement area = 2800 mm^2

Then

$$\phi P_0 = 0.7[0.85 \times 21(62,500 - 2800) + 2800 \times 400]/1000$$
$$= 1530 \text{ kN}$$

To account for accidental eccentricity, ACI Section 10.3.5 requires a spirally reinforced column to be designed for a minimum eccentricity of approximately 0.05h, which gives a maximum design axial load strength at zero eccentricity, in accordance with ACI Equation (10-1), of

$$\phi P_{n\max} = 0.85 \phi P_0$$

In the case of a column with lateral ties, a minimum eccentricity of approximately 0.10h is specified, which gives a maximum design axial load strength at zero eccentricity, in accordance with ACI Equation (10-2), of

$$\phi P_{n\max} = 0.80 \phi P_0 = 0.80 \times 1530 = 1224 \text{ kN}$$

REFERENCES

1. Newnan, D. G., *Civil Engineering License Review.* Engineering Press, Austin, TX, 1995.
2. American Concrete Institute. *Building Code Requirements and Commentary for Reinforced Concrete* (ACI 318-89). Detroit, MI, 1995.
3. American Concrete Institute. *Design Handbook in Accordance with the Strength Design Method.* Detroit, MI, 1985.
4. Ghosh, S. K., and Domel, A. W. *Design of Concrete Buildings for Earthquake and Wind Forces.* Portland Cement Association, Skokie, IL, 1992.
5. Williams, A. *Design of Reinforced Concrete Structures.* Engineering Press, Austin, TX, 1996.

PROBLEMS

In each of the following problems, the short column shown in Exhibit 4.1 supports a steel column and base plate with the indicated loads. The short column is 450 mm square and is reinforced with Grade 400M deformed bars. Concrete compressive strength is 14 MPa. The short column is subjected to axial load only and is effectively braced against sidesway by the floor slab.

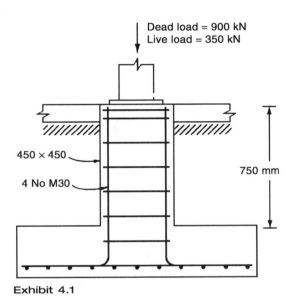

Exhibit 4.1

4.1 The design axial load strength of the short column is given most nearly by
a. 1360 kN   c. 1760 kN
b. 1560 kN   d. 1960 kN

4.2 The factored applied load is given most nearly by
a. 1850 kN   c. 1950 kN
b. 1900 kN   d. 2000 kN

4.3 The minimum allowable reinforcement area in the short column is most nearly
a. 2000 mm^2 c. 2200 mm^2
b. 2100 mm^2 d. 2300 mm^2

SOLUTIONS

4.1 d. For a column with lateral ties, the design axial load strength is given by Eq. (10.2) as

$$\phi P_n = 0.80 \ \phi[0.85 f_c'(A_g - A_{st}) + f_y A_{st}]$$
$$= 0.8 \times 0.7[0.85 \times 14(202{,}500 - 2800) + 400 \times 2800]/1000$$
$$= 1958 \text{ kN}$$

4.2 a. The applied ultimate load is

$$P_u = 1.4 \times 900 + 1.7 \times 350$$
$$= 1855 \text{ kN}$$

4.3   a.   The minimum allowable reinforcement area, in accordance with ACI Section 10.9.1, is

$$\begin{aligned}
\rho_{\min} &= 0.01\, A_g \\
&= 0.01 \times 202{,}500 \\
&= 2025 \text{ mm}^2
\end{aligned}$$

CHAPTER 5

Wastewater and Solid Waste Treatment

Kenneth J. Williamson

WASTEWATER FLOWS

Wastewater flows comprise domestic and industrial wastewaters, infiltration, inflow, and storm water. Most modern sanitary sewers are separated from storm water systems, so these flows are treated separately. Modern sewers are constructed so that inflow rates are assumed to be negligible.

Domestic flows are determined from water use rates, with typical values being 0.57 m^3 per day per capita. Industrial and infiltration flows vary widely. Specific data are required based upon industry and production rates.

SEWER DESIGN

Hydraulics of Sewers

Sewers are designed as open channels, usually with a circular cross section. Flows in sewers are modeled using Manning's equation (see Chapter 3) as

$$V = \frac{1}{n} R^{2/3} S^{1/2} \tag{5.1}$$

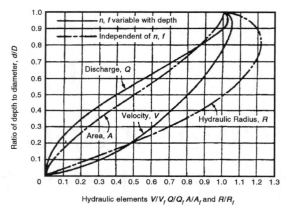

Figure 5.1

where
    V = velocity (m/s)
    n = Manning coefficient, 0.013
    R = hydraulic radius (m)
    S = slope of hydraulic grade line

The relation between the hydraulic radius and the flow depth for a circular cross section is a complex relationship; as a result, flows in partially full sewers are calculated using nomographs like that in Fig. 5.1.

Design calculations usually involve knowing the slope of the sewer (S) and the partially full flow rate, Q. The nomograph is solved by assuming a pipe diameter (D), solving for the full flow rate, Q_f. From the ratio of Q/Q_f, the ratio of partial depth to pipe diameter (d/D) is obtained from the nomograph, from which the depth of flow is calculated. Sanitary sewers are designed to carry the peak flow with a depth of flow from one-half to full.

WASTEWATER CHARACTERISTICS

Wastewater can be characterized by a large number of parameters. Typical concentrations for design are given in Table 5.1. Per capita loading rates for 5-day biochemical oxygen demand and total suspended solids are 100 and 120 g/capita-d, respectively.

Table 5.1 Typical composition of domestic wastewater

Constituent	Concentration (mg/L)
Dissolved solids	700
Volatile dissolved solids	300
Total suspended solids	220
Volatile suspended solids	135
5-day biochemical oxygen demand	200
Organic nitrogen	15
Ammonia nitrogen	25
Total phosphorus	8

Oxygen Demand

Theoretical Oxygen Demand

Theoretical oxygen demand (ThOD) is a value calculated as the oxygen required to convert organic compounds in the wastewater to carbon dioxide and water. To determine the ThOD, the chemical formula of the waste must be known.

Chemical Oxygen Demand

Chemical oxygen demand (COD) is an empirical parameter representing the equivalent amount of oxygen that would be used to oxidize organic compounds under strong chemical oxidizing conditions. The COD and the ThOD are approximately equal for most wastes.

Biochemical Oxygen Demand

Biochemical oxygen demand (BOD) is an empirical parameter representing the amount of oxygen that would be used to oxidize organic compounds by aerobic bacteria. The test involves seeding of diluted wastewater and measurement of the oxygen depletion after five days in 300-mL glass bottles. The 5-day BOD (BOD_5) is related to the ultimate BOD as

$$BOD_5 = BOD_L(1 - 10^{-kt}) \tag{5.2}$$

where
BOD_L = the maximum BOD exerted after a long time of incubation (mg/L);
k = the base-10 BOD decay coefficient (d^{-1})

The BOD test is accomplished at 20°C; the decay coefficient can be converted to other temperatures as

$$k_T = k_{20}\theta^{(T-20)} \tag{5.3}$$

where
k_T, k_{20} = BOD decay coefficient at temperature T and 20°C, respectively
θ = temperature correction coefficient, 1.056 for 20–30°C and 1.135 for 4–20°C

Nitrogenous Oxygen Demand

Nitrogenous oxygen demand results from the oxidation of ammonia to nitrate by nitrifying bacteria. The overall equation is

$$NH_4^+ + 2O_2 = NO_3^- + 2H^+ + 2H_2O \tag{5.4}$$

Nitrogenous oxygen demand is not included in either the BOD_5 or the COD value.

WASTEWATER TREATMENT

Wastewater treatment in the United States is managed under the Federal Water Pollution Control Act Amendments of 1972. Design of wastewater treatment facilities focuses on the two parameters BOD and total suspended solids (TSS). The general approach is to link together a series of unit processes to remove BOD and TSS sequentially to meet the discharge requirements. The unit processes are typically chosen to minimize treatment costs.

Process Analysis

Types of Reactions

Most reactions in wastewater treatment are considered to be homogeneous. In homogeneous reactions, the reaction occurs throughout the liquid, and mass transfer effects can be ignored. Chlorination is an example of a homogeneous reaction. In a heterogeneous reaction, the reactants must be transferred to a reactive site and the products must be transferred away from the reactive site. Biological reactions in bacterial films are an example of heterogeneous reactions.

Reaction Rates

Chemical reactions have many different forms, although the most common form is the conversion of a single product to a single reactant:

$$A \rightarrow B \qquad (5.5)$$

If the rate of the reaction is zero-order, then the rate of conversion of A is

$$\frac{dA}{dt} = -k_0 \qquad (5.6)$$

where k_0 = the zero-order reaction coefficient (mg/L-d).

If the rate of the reaction is first-order, then the rate of conversion of A is

$$\frac{dA}{dt} = -k_1[A] \qquad (5.7)$$

where k_1 = first-order reaction coefficient (d^{-1}).

Reactor Types

Three important reactor types are batch, complete-mix, and plug flow, as shown in Fig. 5.2. Batch reactors have no influent or effluent and have a hydraulic detention time of t_b, which is the time between filling and emptying. Complete-mix reactors have constant influent and effluent flow rates and are mixed so that spatial gradients of concentration are near zero. They have a hydraulic detention time as

Batch

$$\theta_h = \frac{V}{Q} \qquad (5.8)$$

where
   θ = hydraulic detention time (d)
   V = reactor volume (L)
   Q = influent and effluent flow rate (L/d)

Complete mix

Plug flow reactors have constant influent and effluent flow rates but lack internal mixing. The reactors tend to be long and narrow, and the transport of water occurs from one end to the other. Their hydraulic detention time is

Plug-flow

Figure 5.2

$$t_r = \frac{L}{v_e} \qquad (5.9)$$

where
   t_r = hydraulic detention time (d)
   L = length of reaction (m)
   v_e = liquid flow velocity (m/d)

The application of mass balances to the three types of reactors, assuming either zero- or first-order reaction rates, results in the following equations for the effluent concentration of A at steady state:

Batch Reactor, Zero-Order

$$A = A_0 - k_0 t \tag{5.10}$$

Complete-Mix, Zero-Order

$$A = A_0 - k_0 \theta_h \tag{5.11}$$

Plug Flow, Zero-Order

$$A = A_0 - k_0 t_r \tag{5.12}$$

Batch Reactor, First-Order

$$A = A_0 e^{(-k_1 t_b)} \tag{5.13}$$

Complete-Mix, First-Order

$$A = \frac{A_0}{1 + k_1 \theta_k} \tag{5.14}$$

Plug Flow, First-Order

$$A = A_0 e^{-k_1 t_r} \tag{5.15}$$

Plug flow reactors will consistently give lower effluent concentrations as compared to complete-mix reactors for all reaction rates greater than zero.

Physical Treatment Processes

Sedimentation
Sedimentation is divided into four types: discrete (Type 1); flocculent (Type 2); zone (Type 3); and compression (Type 4).

Discrete Sedimentation
Discrete sedimentation is described in Chapter 7. In discrete sedimentation, the removal is determined by the terminal settling velocity. Settling velocities for small Reynolds numbers can be estimated by Stokes' law as

$$V_p = \frac{g(\rho_s - \rho_w)d^2}{18\mu} \tag{5.16}$$

where
V_p = particle settling velocity (m/s)
g = gravitational constant (9.8 m/s^2)
ρ_p, ρ_w = density of particle and water, respectively (kg/m^3)
μ = water viscosity (N-s/m^2)

For particles with V_p less than the overflow rate of the sedimentation basin, V_c, the removal R is

$$R = \frac{V_p}{V_c} \qquad (5.17)$$

R is 1.0 for all particles with $V_p > V_s$.

Flocculent Sedimentation

Flocculent sedimentation is similar to Type 1 settling except that the particle size increases as the particle settles in the sedimentation reactor. Type 2 or flocculent settling occurs in coagulation/flocculation treatment for water and in the top of secondary sedimentation basins.

Removal under flocculent sedimentation is determined empirically. Data must be collected to give isolines for percent removal as shown in Fig. 5.3. Given such data, the total removal is

$$R = \sum \left(\frac{\Delta h_n}{h_s} \right) \left(\frac{R_n + R_{n+1}}{2} \right) \qquad (5.18)$$

where the symbols are as illustrated in Fig. 5.3.

Zone Settling

Zone settling occurs in the bottom of a secondary sedimentation basin, where the concentration of the solids increases to over 5000 mg/L. A schematic of a sedimentation basin coupled to a biological treatment reactor with cell recycle is shown in Fig. 5.4. Under such conditions, the modeling of the zone sedimentation process is based upon the solids flux through the bottom section of the basin. The solids flux results from two components: gravity sedimentation and liquid velocity from the underflow.

$$\text{SF} = \text{SF}_g + \text{SF}_r \qquad (5.19)$$

where

    SF, SF_g, SF_r = solids flux, solids flux due to gravity, and solids flux due to recycle, respectively (kg/m^2-d)

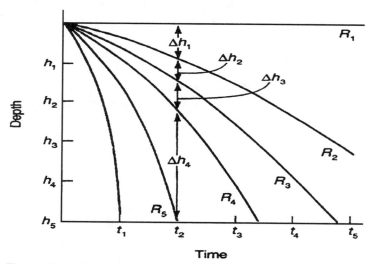

Figure 5.3   Empirical analysis of Type 3 settling

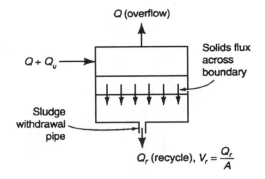

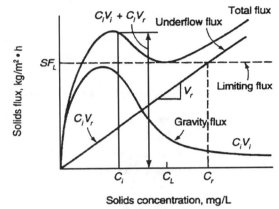

Figure 5.4 Schematic of secondary clarifier and empirical analysis of Type 3 settling

The resulting relationship is

$$SF = C_iV_i + C_iV_r \qquad (5.20)$$

where

C_i = concentration of solids at a specified height in the basin (mg/L)
V_i = gravity settling velocity of sludge with concentration C_i (m/d)
V_r = downward fluid velocity from recycle (m/d)
   $= Q_r/A_H$
A_H = horizontal area of sedimentation basin

Equation (5.19) is graphed in Fig. 5.4. Under a constant recycle flow and a defined sludge, the transport of solids through the bottom of the clarifier becomes limited by the limiting solids flux rate (SF_L) shown in Fig. 5.4. Estimates of SF_L are obtained by graphical analysis as illustrated in Fig. 5.4. The horizontal area of the sedimentation basin is then calculated as

$$A = \frac{(Q + Q_r)(C_{\text{inf1}})}{SF_L} \qquad (5.21)$$

or

$$A = \frac{Q_r C_r}{SF_L} \qquad (5.22)$$

where

C_{inf1} = concentration of TSS in influent to sedimentation basin
C_r = concentration of TSS in recycle from sedimentation basin
Q, Q_r = plant influent and recycle flow rates

Compression Sedimentation

Compression sedimentation involves the slow movement of water through the pores in a sludge cake. The process is modeled as an exponential decrease in height of the sludge cake as

$$H_t - H_\infty = (H_0 - H_\infty)e^{-i(t-t_0)} \tag{5.23}$$

where

H_t, H_∞, H_0 = sludge height after time t, a long period, and initially
i = compression coefficient.

Biological Treatment Processes

Biological treatment involves the conversion of organic and inorganic compounds by bacteria with subsequent growth of the organisms. The common biological treatment processes in wastewater treatment involve the removal of various substrates including BOD, ammonia, nitrate, and organic sludges.

Description of Homogeneous Processes

The rate of removal of a substrate is given as

$$r_{\text{su}} = -\frac{k \times S}{(K_s + S)} \tag{5.24}$$

where

r_{su} = rate of substrate removal (mg/L-d)
k = maximum substrate removal rate (mg/mg-d)
S, K_s = substrate and half-velocity coefficient, respectively (mg/L)

The growth of organisms is given as

$$r_X = Yr_{\text{su}} - k_d X \tag{5.25}$$

where

r_X = rate of growth of bacteria (mg/L-d)
Y = substrate yield rate (mg/mg)
k_d = bacterial decay coefficient (d^{-1})
X = bacterial concentration, usually expressed as TVSS (mg/L)

Complete-Mix Reactor Without Cell Recycle

For the complete-mix reactor without cell recycle shown in Fig. 5.5(a), the governing equations are

$$X = \frac{Y(S_0 - S)}{(1 + k_d \theta_h)} \tag{5.26}$$

$$S = \frac{K_s(1 + \theta_h k_d)}{\theta_h(YK - k_d) - 1} \tag{5.27}$$

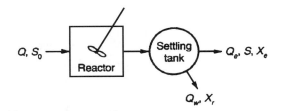

(a) Complete-Mix Without Cell Recycle

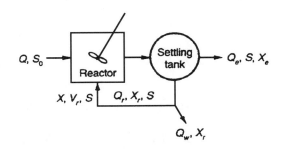

(b) Complete-Mix With Cell Recycle

Figure 5.5 Schematic of complete-mix
reactors

Complete-Mix Reactor with Cell Recycle
For the complete-mix reactor with cell recycle shown in Fig. 5.5(b), the governing
equations are

$$\frac{1}{\theta_c} = Y \left(\frac{kSX}{S+K_s} \right) - k_d \tag{5.28}$$

where θ_c = solids retention time (d);

$$\frac{F}{M} = \frac{S_0}{\theta_h X} \tag{5.29}$$

where F/M = food-to-microorganism ratio (mg/mg-d); and

$$X = \frac{\theta_c Y (S_0 - S)}{\theta_h (1 + k_d \theta_c)} \tag{5.30}$$

$$S = \frac{K_s (1 + \theta_c k_d)}{\theta_c (Yk - k_d) - 1} \tag{5.31}$$

Process design proceeds from selection of a solids retention time (θ_c) as the
controlling parameter. The volume of the reactor is determined from Eq. (5.30)
with $\theta_h = V/Q$. The waste sludge flow rate is computed from

$$\theta_c = \frac{VX}{Q_w X_w + Q_e X_e} \tag{5.32}$$

where

Q_w, Q_e = sludge waste and effluent flow rates (L/d)
X_w = sludge waste concentration (X if from reaction, X_r if from the clarifier) (mg/L)
X_e = effluent TSS concentration (mg/L), usually 20 mg/L

Plug Flow Reactor with Cell Recycle
The cell concentration in a plug flow reactor can be estimated using Eq. (5.30). The effluent substrate concentration can be estimated from

$$\frac{1}{\theta_c} = \frac{Yk(S_0 - S)}{(S_0 - S) + \left(1 + \frac{Q_r}{Q}\right)K_s \ln(S_i/S)} - k_d \tag{5.33}$$

where S_i = the diluted substrate concentration entering the reactor (mg/L).

Description of Heterogeneous Processes

Trickling filters are designed using an empirical equation as

$$A = Q\left(\frac{-\ln(S_{eff}/S_{infl})}{k_1 D}\right)^2 \tag{5.34}$$

where

A = horizontal area
k_1 = BOD decay coefficient
D = filter depth

Disinfection

Disinfection in wastewater treatment is almost universally accomplished using chlorine gas. The objective of disinfection is to kill bacteria, viruses, and amoebic cysts. Effluent standards for secondary treatment require fecal coliform levels of less than 200 and 400 per 100 mL for 30-day and 7-day averages, respectively.

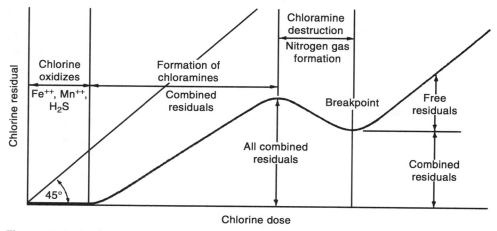

Figure 5.6 Chlorine dosage versus chlorine residual

Modeling of disinfection is described in Chapter 7. The major difference between chlorination of water supplies and that of wastewater is the reaction of chlorine with ammonia to produce chloramines as

$$NH_3 + HOCl \rightarrow NH_2Cl + H_2O$$

Chloramines can undergo further oxidation to nitrogen gas with their subsequent removal.

The stepwise oxidation of various compounds is shown in Fig. 5.6. As chlorine is initially added, it reacts with easily oxidized compounds such as reduced iron, sulfur, and manganese. Further addition results in the formation of chloramines, collectively termed **combined chlorine residual**. Further addition of chlorine results in destruction of the chloramines and ultimately to the formation of free chlorine residuals as HOCl and OCl⁻. The development of free chlorine residual is termed **breakpoint chlorination**.

PROBLEMS

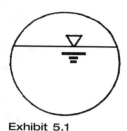

Exhibit 5.1

5.1   What is the required slope for a 0.3-m diameter circular sewer as shown in Exhibit 5.1, with a flow of 0.014 m^3/s and a minimum velocity of 0.6 m/s?

5.2   Compute the carbonaceous and nitrogenous oxygen demands for a waste with a chemical formula of $C_5H_7NO_2$.

5.3   A lake with a volume of 5×10^6 m^3 has a freshwater flow of 20 m^3/s. A waste is dumped into the lake at a rate of 50 g/s with a decay rate of 0.2/d. What is the steady-state concentration? Assume that the lake is completely mixed.

5.4   A waste with a flow of 2.8 L/s is discharged to a small stream with a flow of 141 L/s. The waste has a 5-d BOD of 200 mg/L ($k = 0.2$/d). What is the BOD after 1 day's travel in the stream?

5.5   A sedimentation basin has an overflow of 3 ft/hr. The influent wastewater has a particle distribution as follows:

Percent of Particles	Settling Velocity (m/hr)
20	0.30 to 0.61
30	0.61 to 0.91
50	0.91 to 1.22

Determine the total removal in the sedimentation basin.

5.6   Disinfection with chlorine is known to be a first-order reaction. The first-order decay rate under a given concentration of chlorine is measured as 0.35/hr. The flow rate is 12,000 L per hour, and the desired removal of organisms is from 10×10^6/100mL to less than 1/100mL. Determine
   a.   Volume of reactor required assuming the use of a complete-mix reactor
   b.   Volume of reactor required assuming the use of plug flow reactor

5.7   A Type 3 settling test was conducted on a waste activated sludge. The results were as follows:

MLSS (mg/L)	Settling Velocity (m/hr)
4000	2.4
6000	1.2
8000	0.55
10,000	0.31
20,000	0.06

Determine the limiting solids flux if the concentration of the recycled solids is 10,000 mg/L.

5.8 Find the effluent soluble BOD and reactor cell concentration for an aerobic complete-mix reactor with no recycle.

$$k = 10 \text{ mg/mg-d}$$
$$K_s = 50 \text{ mg/L}$$
$$k_d = 0.10\text{/d}$$
$$Y = 0.6 \text{ mg/mg}$$
$$S_0 = 200 \text{ mg BOD/L}$$
$$\theta_h = \theta_c = 2\text{d}$$

SOLUTIONS

5.1
$$D_f = 0.30 \text{ m}$$
$$Q = 0.014 \text{ m}^3\text{/s}$$
$$V = 0.6 \text{ m/s}$$

Need V/V_f or Q/Q_p or A/A_f or R/R_f

$$A_f = \frac{\pi D_f^2}{4} = 0.071 \text{ m}^2$$

$$A = \frac{Q}{V} = 0.023 \text{ m}^2$$

$$\frac{A}{A_f} = 0.32$$

From nomograph (Exhibit 5.1a)

$$d/D_f = 0.35 \quad d = 0.105 \text{ m}$$
$$R/R_f = 0.75$$

$$R_f = \frac{\pi D_f^2/4}{\pi D} = \frac{D_f}{4} = 0.075 \text{ m}$$

$$R = (0.75)(0.075) = 0.056 \text{ m}$$

From Manning's equation:

$$V = \frac{1}{n} R^{2/3} S^{1/2}$$

$$S = \left(\frac{nV}{1.R^{2/3}} \right)^2$$

$$= \left(\frac{(0.013)(0.6 \text{ m/s})}{(1.)(0.056)^{2/3}} \right)^2$$

$$= 0.0028 \text{ m/m}$$

5.2 For ThOD:

$$C_5H_7NO_2 + 5O_2 \rightarrow 5CO_2 + NH_3 + 2H_2O$$

For NOD:

$$NH_3 + 2O_2 \rightarrow HNO_3 + H_2O$$

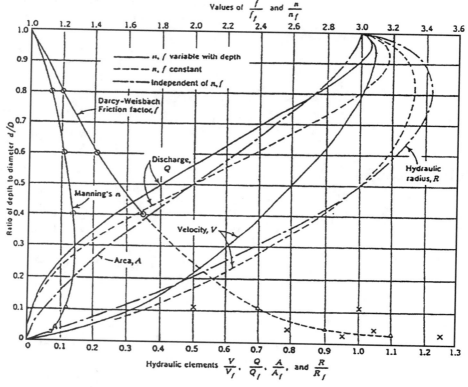

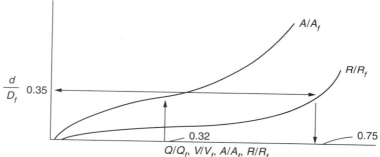

Exhibit 5.1a

For carbonaceous oxygen demand:

$$\frac{5 \text{ moles O}_2}{1 \text{ mole C}_5\text{H}_7\text{NO}_2} \times \frac{32 \text{ g O}_2}{\text{mole O}_2} \times \frac{1 \text{ mole C}_5\text{H}_7\text{NO}_2}{113 \text{ g}} = \frac{1.4 \text{ g O}_2}{\text{g C}_5\text{H}_7\text{NO}_2}$$

For nitrogenous oxygen demand:

$$\frac{2 \text{ moles O}_2}{1 \text{ mole C}_5\text{H}_7\text{NO}_2} \times \frac{32 \text{ g O}_2}{\text{mole O}_2} \times \frac{1 \text{ mole C}_5\text{H}_7\text{NO}_2}{113 \text{ g}} = \frac{0.57 \text{ g O}_2}{\text{g C}_5\text{H}_7\text{NO}_2}$$

5.3

$$C = \frac{S}{Q + kV}$$

$$= \frac{50 \text{ g/s}}{20 \text{ m}^3/\text{s} + (0.2/\text{d})(5 \times 10^6 \text{ m}^3)\left(\dfrac{1 \text{ d}}{8.64 \times 10^4 \text{ s}}\right)}$$

$$= 1.58 \text{ g/m}^3$$

5.4
$$BOD_L = \frac{BOD_5}{(1+e^{-5k})}$$

$$= \frac{20 \text{ mg/L}}{\left(1-e^{-\left(5d\times\frac{2}{d}\right)}\right)}$$

$$= 294 \text{ mg/L}$$

After mixing,

$$BOD_i = \frac{(294 \text{ mg/L})(2.8 \text{ L/s})}{141.6 \text{ L/s} + 2.8 \text{ L/s}}$$
$$= 5.76 \text{ mg/L}$$

Assume the stream is plug flow.

$$BOD = BOD_i e^{-kt}$$
$$= (5.76 \text{ mg/L})e^{-(0.21d)(1d)}$$
$$= 4.7 \text{ mg/L}$$

5.5

Particles	Avg V_p (cm/hr)	V_c (cm/hr)	R (%)
2.54–6.45	3.81	7.62	50
6.45–7.62	6.35	7.62	83
7.62–10.16	8.89	7.62	100

$$R_{total} = (0.2)(0.5) + (0.3)(0.83) + (0.5)(1.0)$$
$$= 0.85 \text{ or } 85\%$$

5.6

a.
$$\frac{A}{A_0} = \frac{1}{1+k_1\frac{v}{Q}} = \frac{1}{10\times10^6}$$

$$1+k_1\frac{v}{Q} = 1\times10^7$$

$$V = \frac{(1\times10^7)(12,000 \text{ L/hr})}{0.35/\text{hr}}$$
$$= 3.4\times10^{11} \text{ L}$$

b.
$$\frac{A}{A_0} = e^{-kv/Q} = \frac{1}{10\times10^6}$$

$$-k\frac{v}{Q} = -16.1$$

$$V = \frac{-16.1(12,000 \text{ L/hr})}{0.35/\text{hr}}$$
$$= 5.5\times10^5 \text{ L}$$

5.7

MLSS (mg/L)	v_i (ft/hr)	$x_i v_i$ (mg-ft/L-hr)	v_i (cm/hr)	$x_i v_i$ (kg/m²-d)
4000	7.8	3.12×10^4	19.81	4.63
6000	3.8	2.28×10^4	9.65	3.11
8000	1.8	1.44×10^4	4.57	1.97
10,000	1.0	1.00×10^4	2.54	1.37
20,000	0.2	0.80×10^4	0.51	1.09

From graphical solution (see Exhibit 5.7)

Limiting solids flux = 7 kg/m²-d

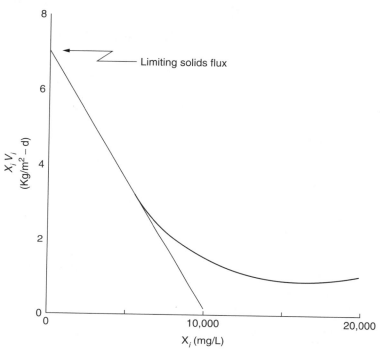

Exhibit 5.7

5.8

$$s = \frac{K_s(1+\theta_h k_d)}{\theta_h(Y_k - k_d)-1}$$

$$= \frac{(50 \text{ mg/L})(1+(2 \text{ d})(0.10/\text{d}))}{(2 \text{ d})\left(\dfrac{0.6 \text{ mg}}{\text{mg}} \times \dfrac{10 \text{ mg}}{\text{mg-d}} - \dfrac{0.10}{\text{d}}\right)^{-1}}$$

$$= 5.7 \text{ mg/L}$$

$$X = \frac{4(S_0 - S)}{1+k_d\theta_h} = \frac{\left(\dfrac{0.6 \text{ mg}}{\text{mg}}\right)(200 \text{ mg/L} - 5.7 \text{ mg/L})}{1+(0.10/\text{d})(2 \text{ d})}$$

$$= 97 \text{ mg/L}$$

Transportation Engineering

Robert W. Stokes

HIGHWAY CURVES

Simple (Circular) Horizontal Curves

The location of highway centerlines is initially laid out as a series of straight lines (tangent sections). These tangents are then joined by circular curves to allow for smooth vehicle operations at the design speed selected for the highway. Figure 6.1 shows the basic geometry of a simple circular curve. If the two tangents intersecting at PI are laid out, and the angle Δ between them is measured, only one other element of the curve need be known to calculate the remaining elements. The radius of the curve (R) is the other element most commonly used.

Circular Curve Formulas
The following symbols are defined as shown in Fig. 6.1:

    PC = Point of curvature (beginning of curve)
     PI = Point of intersection
    PT = Point of tangency (end of curve)
     Δ = Intersection or central angle, degrees

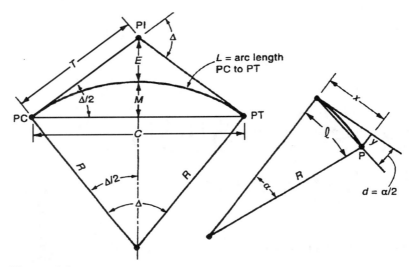

Figure 6.1

R = Radius of curve, m
L = Length of curve, m
E = External distance, m
M = Middle ordinate, m
C = Long chord, m
P = Length of arc between any two points on curve, m
α = Central angle subtended by arc P, degrees
d = Deflection angle for any arc length P, degrees
x = Distance along tangent from PC or PT to set any point P on curve, m
y = Offset (normal) from tangent at distance x to set any point P on curve, m

Then

$L = \Delta R/57.2958$
$T = R \tan \Delta/2$
$E = R(\sec \Delta/2 - 1) = R \operatorname{exsec} \Delta/2$
$M = R(1 - \cos \Delta/2) = R \operatorname{vers} \Delta/2$
$C = 2R \sin \Delta/2$
$R = (C^2 + 4M^2)/8M$ (formula for estimating R from field measurements)
$P = (\alpha \times R)/57.2958$
$d = \alpha/2 = 1718.873 \; P/R$ (in minutes)
   $= 28.64789 \; P/R$ (in degrees)

For any length x,

$$y = R - (R^2 - x^2)^{1/2}$$

For any length P,

$$X = R \sin \alpha$$
$$Y = R (1 - \cos \alpha) = R \operatorname{vers} \alpha$$

Example **6.1**

Given the following horizontal curve data, determine the curve radius, R; length of curve, L; stationing (sta) of the PC; stationing of the PT; the long chord, C; and deflection angle, d, of a point on the curve 30 m ahead of the PC.

$$PI = \text{sta } 2 + 170.00 \text{ (km)}$$
$$\Delta = 41°10'$$
$$T = 115.00 \text{ m}$$

Solution

$$PC \text{ sta} = PI \text{ sta} - T = 2170.00 - 115.00 = 2 + 055.00$$
$$R = T/\tan(\Delta/2) = 115.00/0.3755 = 306.22 \text{ m}$$
$$L = (\Delta R)/57.2958 = 220.02 \text{ m}$$
$$PT \text{ sta} = PC \text{ sta} + L = 2055.00 + 220.02 = 2275.02 \text{ m}$$
$$C = 2R \sin(\Delta/2) = 2(306.22)(0.3516) = 215.32 \text{ m}$$
$$d = (28.64789P)/R = [(28.64689)(30)]/306.22 = 2.81°$$

Horizontal Curve Design

The minimum radius of horizontal curvature is determined by the dynamics of vehicle operation and sight distance requirements. The minimum radius necessary for the vehicle to remain in equilibrium with respect to the incline of a horizontal curve is given by the following formula:

$$R = V^2/[127 \, (e + f)]$$

where V = vehicle speed in km/h, e = rate of superelevation ("banking") in m/m of width, and f = allowable side friction factor. The equation can be used to solve for the minimum radius for a given speed, or the maximum safe speed for a given radius.

Example **6.2**

An existing horizontal curve has a radius of 235 m. What is the maximum safe speed on the curve? Assume $e = 0.08$ and $f = 0.14$.

Solution

$$R = V^2/[127 \, (e + f)]$$
$$V = [R[127 \, (e + f)]]^{0.5}$$
$$V = [235 \, [127 \, (0.08 + 0.14]]^{0.5} = 81 \text{ km/h}$$

Vertical Curves

The vertical-axis parabola (see Fig. 6.2) is the geometric curve most commonly used in the design of vertical highway curves. The general equation of the parabola is

$$Y = aX^2 + bX + c$$

where the constant a is an indication of the rate of change of slope, b is the slope of the back tangent (g_1), and c is the elevation of the PVC.

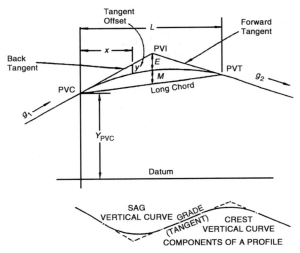

Figure 6.2

Vertical Curve Formulas

The following notation is defined as shown in Fig. 6.2:

$$L = \text{Length of curve (horizontal)}$$
$$\text{PVC} = \text{Point of vertical curvature}$$
$$\text{PVI} = \text{Point of vertical intersection}$$
$$\text{PVT} = \text{Point of vertical tangency}$$
$$g_1 = \text{Grade of back tangent (decimal)}$$
$$g_2 = \text{Grade of forward tangent (decimal)}$$
$$a = \text{Parabola constant}$$
$$y = \text{Tangent offset}$$
$$E = \text{Tangent offset at PVI}$$
$$M = \text{Middle ordinate}$$
$$r = \text{Rate of change of grade}$$
$$K = \text{Rate of curvature}$$
$$A = \text{Algebraic difference in grades}$$
$$x = \text{Horizontal distance from PVC to point on curve}$$
$$x_m = \text{Horizontal distance to min/max elevation on curve}$$

Then

$$A = g_2 - g_1$$
$$a = (g_2 - g_1)/2L = A/2L$$
$$r = (g_2 - g_1)/L = A/L$$
$$K = L/A$$
$$E = a\,(L/2)^2 = (AL)/8 = M$$
$$y = ax^2$$
$$x_m = -g_1/(2a) = -g_1 K$$

Tangent elevation $= Y_{\text{PVC}} + g_1 x$

Curve elevation $= Y_{\text{PVC}} + g_1 x \pm ax^2$ (add for sag curves, subtract for crest curves)

Example 6.3

Determine the following for the sag vertical curve shown in Exhibit 1

a. Stationing and elevation of the PVC

b. Stationing and elevation of the PVT

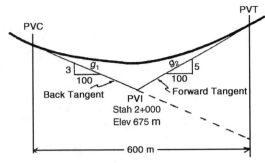

Exhibit 1

c. Elevation of the curve 100 m from the PVC

d. Elevation of the curve 400 m from the PVC

e. Stationing and elevation of the low point of the curve

Solution

a. The curve is symmetric about the PVI, therefore:

Sta PVC = Sta PVI – $L/2$ = 2000 – 600/2 = 1 + 700.0
Elev of the PVC = Elev PVI + $g_1(L/2)$ = 675 + 0.03(300) = 684.0 m

b. Sta PVT = Sta PVI + $L/2$ = 2000 + 600/3 = 2 + 300.0

Elev of the PVT = Elev PVI + $g_2(L/2)$ = 675 + 0.05(300) = 690.0 m

c. The elevation of the curve at any distance x from the PVC is equal to the tangent offset + the tangent elevation. To calculate the tangent offset, it is first necessary to evaluate the constant a.

$$a = (g_2 - g_1)/2L = [0.05 - (-0.03)]/[2(600)] = 0.000067$$

Tangent offset at x = 100 m from the PVC = ax^2 = 0.000067 $(100)^2$ = 0.67 m
Tangent elevation 100 m from the PVC = PVC elev + g_1x
= 684.0 m + (–0.03)100 = 681.0 m

Curve elevation = tangent offset + tangent elev = 0.67 + 681.0 = 681.67 m

d. The elevation of the curve 400 m from the PVC can be calculated relative to the forward tangent or relative to the extension of the back tangent. Relative to the extension of the back tangent:

Tangent offset = ax^2 = 0.000067 $(400)^2$ = 10.67 m
Elev of the extended back tangent = PVC elev + g_1x
= 684.0 + (–0.03)400 = 672.0 m
Curve elev = tangent offset + tangent elev = 10.67 + 672.0 = 682.67 m

e. The low point of the curve occurs at a distance x_m from the PVC.

$x_m = -g_1/(2a) = -(-0.03)/[2(0.000067)] = 225$ m (from the PVC)

Sta of the low point = Sta PVC + 225 = 1700 + 225 = 1 + 925.0
The tangent offset at 225 m from the PVC = ax^2 = 0.000067$(225)^2$ = 3.38 m
The tangent elevation at this distance = PVC elev + g_1x
= 684.0 + (–0.03)225 = 677.25 m
The elevation of the low point = tangent offset + tangent elevation
= 677.25 + 3.38 = 680.63 m

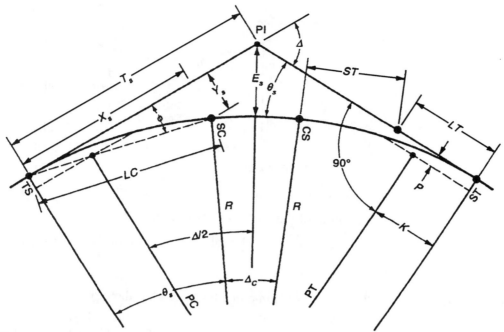

Figure 6.3

Transition (Spiral) Curves

The spiral curve is used to allow for a transitional path from tangent to circular curve, from circular curve to tangent, or from one curve to another that have substantially different radii. The minimum length of spiral curve needed to achieve this transition can be computed from the following formula.

$$P_s = V^3/(46.7RC)$$

where P_s = minimum length of spiral (m), V = design speed (km/hr), R = circular curve radius (m), and C = rate of increase of centripetal acceleration (m/s^3). The value $C = 0.6$ is generally used for highway curves.

Transition (Spiral) Curve Formulas

The basic geometry of the spiral curve is shown in Fig. 6.3. The spiral is defined by its parameter A and the radius of the simple curve it joins. The product of the radius (r) at any point on the spiral and the corresponding spiral length (P) from the beginning of the spiral to that point is equal to the product of the radius (R) of the simple curve it joins and the total length (P_s) of the spiral as shown in the following equation:

$$rR = RR_s = \text{constant} = A^2$$

The following notation is defined as shown in Fig. 6.3:

ℓ_s = total length of spiral from TS to SC
ℓ = spiral length from TS to any point on spiral
L_c = total length of circular curve
R = radius of circular curve
T_s = total tangent distance from the PI to the TS or ST
Δ = deflection angle between the tangents

θ_s = central angle of the entire spiral (spiral angle)

θ = central angle of any point on the spiral

Δ_c = central angle of simple curve

E_s = external distance

LC = long chord

LT = long tangent

ST = short tangent

X_s = tangent distance from TS to SC

x_s = tangent distance from TS to any point on the spiral

Y_s = tangent offset at the SC

y_s = tangent offset at any point on the spiral

k = simple curve coordinate (abscissa)

p = simple curve coordinate (ordinate)

ϕ = deflection angle at TS from the initial tangent to any point on the spiral

TS = tangent to spiral

SC = spiral to circular curve

CS = circular curve to spiral

ST = spiral to tangent

C_s = spiral deflection angle correction factor

Then

$$\theta_s = 28.648 \ (P_s/R)$$

$$\theta = (P/P_s)^2 \theta_s$$

$$\phi = (\theta/3) - C_s$$

$$C_s \ \text{(seconds)} = 0.0031\theta^3 + 0.0023\theta^5 \times 10^{-5}$$

$$Y_s = (\ell_s^2/6R) - (\ell_s^4/133R^3)$$

$$y_s = (\ell^2/6R) - (\ell^4/336R^3)$$

$$X_s = \ell_s - (\ell_s^3/40R^2)$$

$$p = Y_s - R(1 - \cos\theta_s)$$

$$k = X_s - R\sin\theta_s$$

$$T_s = (R + p)\tan(\Delta/2) + k$$

$$ST = Y_s/\sin\theta_s$$

$$LT = X_s - (Y_s/\tan\theta_s)$$

All distances are in meters and all angles are in degrees unless noted otherwise.

Example 6.4

A transition spiral is being designed to provide a gradual transition into a circular curve with a radius of 435 m and a design speed of 90 km/h. The PI of the curve is at Station 0 + 625.00, and the deflection angle between the tangents is 33.67 degrees. Determine the minimum length of spiral required and the stationing of the TS and the ST.

Solution

Assuming a value of $C = 0.6$, the minimum length of the spiral can be determined as follows.

$$\ell_s = V^3/(46.7RC) = (90)^3/(46.7 \times 435 \times 0.6) = 60 \ \text{m}$$

To determine the stationing of the TS and the ST, proceed as follows:
Compute the spiral angle.

$$\theta_s = 28.648 \ (P_s/R) = 28.648 \ (60/435) = 4 \text{ degrees}$$

Compute Y_s and X_s.

$$Y_s = \left(\ell_s^2/6R\right) - \left(\ell_s^4/336R^3\right) = [(60)^2/(6)(435)] - [(60)^4/(336)(435)^3] = 1.379 \text{ m}$$

$$X_s = \ell_s - \left(\ell_s^3/40R^2\right) = 60 - [(60)^3/(40)(435)^2] = 59.972 \text{ m}$$

Compute p and k.

$$p = Y_s - R(1 - \cos\theta_s) = 1.379 - 435(1 - \cos 4°) = 0.319 \text{ m}$$
$$k = X_s - R \sin\theta_s = 59.972 - (435) \sin 4° = 29.628 \text{ m}$$

Compute the spiral tangent length.

$$T_s = (R + p) \tan (D/2) + k$$
$$= (435 + 0.319) \tan (33.67/2) + 29.628 = 161.349 \text{ m}$$

Compute the central angle of the circular curve.

$$\Delta_c = \Delta - 2\theta_s = 33.67 - 2(4) = 25.67 \text{ degrees}$$

Compute the length of the circular curve.

$$L = (\Delta_c R)/57.2958 = [25.67(435)]57.2958 = 194.891 \text{ m}$$

Compute the stationing of the TS.

$$\text{TS Sta} = \text{PI Sta} - T_s = 0 + 625 - 0 + 161.349 = 0 + 463.651$$

Compute the stationing of the ST.

$$\text{ST Sta} = \text{TS Sta} + L + 2y\ell_s$$
$$= 0 + 463.651 + 0 + 194.891 + 2(60) = 0 + 778.542$$

SIGHT DISTANCE

The ability of drivers to see the road ahead is of utmost importance in the design of highways. This ability to see is referred to as **sight distance** and is defined as the length of highway ahead that is visible to the driver.

Stopping sight distance is the sum of the distance traveled during the perception-reaction time and the distance traveled while braking to a stop. The stopping sight distance can be determined from the following equation.

$$S = 0.278Vt + V^2/[254 \ (f \pm g)]$$

where S = stopping sight distance (m), V = vehicle speed (km/hr), t = perception-reaction time (assumed to be 2.5 s), f = coefficient of friction, and g = grade (*plus* for uphill, *minus* for downhill), expressed as a decimal.

Sight Distance on Simple Horizontal Curves

The required middle ordinates (M) for clear sight areas to satisfy stopping sight distance (S) requirements as a function of the radii (R) of simple horizontal curves can be determined using the following equation. The stopping sight distance is measured along the centerline of the inside lane of the curve.

$$M = R[1 - \cos(28.65 \ S/R)]$$

Example **6.5**

A large outcropping of rock is located at M (middle ordinate) = 10 m from the centerline of the inside lane of a proposed highway curve. The radius of the proposed curve is 120 m. What speed limit would you recommend for this curve? Explain your answer. Assume $f = 0.28$, $e = 0.10$, perception-reaction time = 2.5 sec, and level grade.

Solution

To solve this problem you must determine whether sight distance or the radius of the curve controls the speed.

Determine the available sight distance.

$$M = R[1 - \cos(28.65 \; S/R)]$$
$$S = (R/28.65) \cos^{-1}[(R - M)/R]$$
$$= (120/28.65)\cos^{-1}[(120 - 10)/120] = 98.67 \text{ m}$$

Determine the maximum safe speed on the curve as a function of stopping sight distance.

$$S = 0.278Vt + V^2/[254(f \pm g)]$$
$$98.67 = 0.278V(2.5) + V^2/[254(0.28)]$$
$$V^2 + 49.43V - 7017.41 = 0$$

and from the quadratic equation

$$V = 62.6 \text{ km/hr}$$

Determine the maximum safe speed on the curve as a function of the radius.

$$R = V^2/[127(e + f)]$$
$$V = [127R(e + f)]^{0.5} = [127(120)(0.10 + 0.28)]^{0.5} = 76.1 \text{ km/hr}$$

Because the sight distance is sufficient only for a speed of about 62.6 km/hr, set speed limit at 60 km/hr.

Sight Distance on Vertical Curves

Formulas for stopping and passing sight distances for *vertical curves* are summarized following. The formulas are based on an assumed driver eye height (h_1) of 1070 mm. The height of object (h_2) above the roadway surface is assumed to be 150 mm for stopping sight distance and 1300 mm for passing sight distance.

Formulas for Sight Distance on Vertical Curves
The following terms are defined as shown in Fig. 6.4:

L = Length of vertical curve, m

A = Algebraic difference grades, %

S = Sight distance, m

K = Vertical curvature, L/A

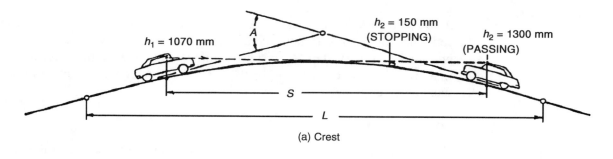

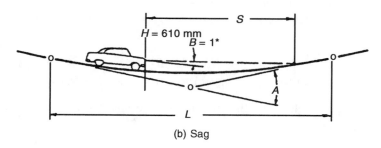

(b) Sag

Figure 6.4

H = Headlight height, m

B = Upward divergence of light beam, degrees

Crest:

$$L = AS^2/100 \left(\sqrt{2h_i} + \sqrt{2h_2} \right)^2 \quad \text{when } S < L$$

The formula is different for when $S > L$, but does not apply to geometric design crieria.

Stopping sight distance: $L = AS^2/404$ or $K = S^2/404$

Passing sight distance: $L = AS^2/946$ or $K = S^2/946$

Sag:

$$L = AS^2/200 \ (H + S \tan B) \quad \text{when } S < L$$

For passenger cars:

$$L = AS^2/(122 + 3.5 \ S) \quad \text{or} \quad K = S^2/(122 + 3.5 \ S)$$

Example 6.6

A crest vertical curve joins a +2% grade with a −2% grade. If the design speed of the highway is 95 km/hr, determine the minimum length of curve required. Assume $f = 0.29$ and the perception-reaction time = 2.5 s. Also assume $S < L$.

Solution

Determine the stopping sight distance required for the design conditions.

$S = 0.278Vt + V^2/[254 \ (f \pm g)] = 0.278 \ (95)(2.5) + (95)^2/[254(0.29 - 0.02)]$
$\quad = 197.62$ m

(Note that the worst-case value for g is used.)

Determine the minimum length of vertical curve to satisfy the required sight distance.

$$L = AS^2/404 = [4(197.62)^2]/404 = 386.67 \text{ m}$$

TRAFFIC CHARACTERISTICS

The traffic on a roadway may be described by three general parameters: volume or rate of flow, speed, and density.

Traffic **volume** is the number of vehicles that pass a point on a highway during a specified time interval. Volumes can be expressed in daily or hourly volumes or in terms of subhourly rates of flow. There are four commonly used measures of daily volume.

1. **Average annual daily traffic** (AADT) is the average 24-hour traffic volume at a specific location over a full year (365 days).

2. **Average annual weekday traffic** (AAWT) is the average 24-hour traffic volume occurring on weekdays at a specific location over a full year.

3. **Average daily traffic** (ADT) is basically an estimate of AADT based on a time period less than a full year.

4. **Average weekday traffic** (AWT) is an estimate of AAWT based on a time period less than a full year.

Highways are generally designed on the basis of the **directional design hourly volume (DDHV)**.

$$\text{DDHV} = \text{AADT} \times K \times D$$

where AADT = average annual daily traffic (veh/day), K = proportion of daily traffic occurring in the design hour, D = proportion of design hour traffic traveling in the peak direction of travel.

For design purposes, K often represents the proportion of AADT occurring during the **thirtieth highest hour** of the year on rural highways, and the fiftieth highest hour of the year on urban highways.

Traffic volumes can also exhibit considerable variation within a given hour. The relationship between hourly volume and the maximum 15-minute rate of flow within the hour is defined as the **peak-hour factor (PHF)**.

$$\text{PHF} = V_h/(4V_{15})$$

where V_h = hourly volume (vph), V_{15} = maximum 15-minute rate of flow within the hour (veh).

The second major traffic stream parameter is **speed**. Two different measures of average speed are commonly used. **Time mean speed** is the average speed of all vehicles passing a point on the highway over a given time period. **Space mean speed** is the average speed obtained by measuring the instantaneous speeds of all vehicles on a section of roadway. Both of these measures can be calculated from a series of measured travel times over a measured distance from the following equations.

$$\mu_t = (\Sigma(d/t_i))/n$$
$$\mu_s = (nd)/\Sigma t_i$$

where μ_t = time mean speed (m/sec or km/hr), μ_s = space mean speed (m/s or km/hr), d = distance traversed (m or km), n = number of travel times observed, t_i = travel time of ith vehicle (s or hr)

Because the space mean speed weights slower vehicles more heavily than time mean speed does, it results in a lower average speed. The relationship between these two mean speeds is

$$\mu_t = \mu_s + \sigma_s^2/\mu_s$$

where σ_s^2 = the variance of the space speed distribution.

Example 6.7

The following travel times were observed for four vehicles traversing a 1-km segment of highway.

Vehicle	Time (minutes)
1	1.6
2	1.2
3	1.5
4	1.7

Calculate the space and time mean speeds of these vehicles.

Solution

The space mean speed is

$$\mu_s = (nd)/\Sigma t_i = 4(1)/(1.6 + 1.2 + 1.5 + 1.7) = 0.67 \text{ km/min} = 40.0 \text{ km/hr}$$

The time mean speed is

$$\mu_t = \Sigma(d/t_i)/n = [(1/1.6) + (1/1.2) + (1/1.5) + (1/1.7)]/4$$
$$= 0.68 \text{ km/min} = 40.8 \text{ km/hr}$$

The third measure of traffic stream conditions, **density** (sometimes referred to as **concentration**), is the number of vehicles traveling over a unit length of highway at a given instant in time.

The general equation relating flow, density, and speed is

$$q = k\mu_s$$

where q = rate of flow (vph), μ_s = space mean speed (km/hr), and k = density (veh/km)

The relationship between density and flow shown in Fig. 6.5 is commonly referred to as the "fundamental diagram of traffic flow." Note in the figure that (1) when density is zero, the flow is also zero; (2) as density increases, the flow also increases; and (3) when density reaches its maximum, referred to as jam density (k_j), the flow is zero.

EARTHWORK

One of the major objectives in evaluating alternative route locations is to minimize the amount of cut and fill. A common method of determining the volume of earth-work is the **average end area** method. This method is based on the assumption that the

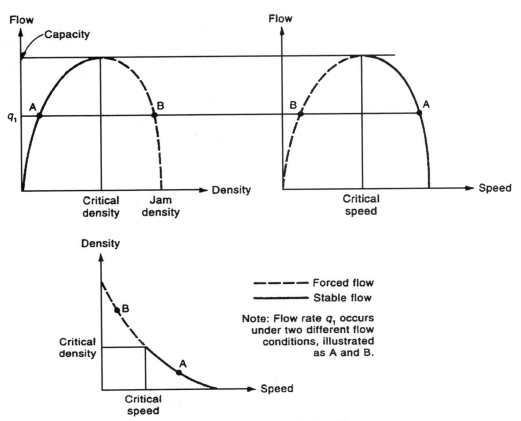

Figure 6.5 Relationships among speed, density, and rate of flow

volume between two consecutive cross sections is the average of their areas multiplied by the distance between them.

$$V = (L/54)(A_1 + A_2)$$

where V = volume (meters^3), A_1 and A_2 = end areas (m^2)

$$L = \text{distance between cross sections (m)}$$

In situations where there is a significant difference between A_1 and A_2, it may be advisable to calculate the volume as a pyramid:

$$V = (1/81)(\text{area of base})(\text{length})$$

where V is in cubic meters, the area of the base is in square meters, and the length is in meters.

The average end area and the pyramid methods provide "reasonably accurate" estimates of volumes of earthwork. When a more precise estimate of volume is desired, the **prismoidal formula** is frequently used:

$$V = (L/6)(A_1 + 4A_m + A_2)$$

where V = volume (m^3), A_1 and A_2 = end areas (m^2), and A_m = middle area determined by averaging corresponding *linear* dimensions (*not* the end areas) of the end sections (m^2).

When materials from cut sections are moved to fill sections, shrinkage factors (generally in the range of 1.10 to 1.25) are applied to the fill volumes to determine the quantities of fill required.

Example 6.8

Given the trapezoidal cross sections shown in Exhibit 2 from a temporary access ramp, determine the volume in cubic meters between stations 4 + 320 and 4 + 400 by (1) the average end area method, and (2) the prismoidal method. Comment on any differences in the results obtained from the two methods.

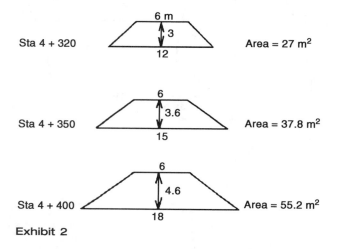

Exhibit 2

Solution

1. For the average end area method:

$$V = L/2 \, (A_1 + A_2)$$

The total volume is the sum of the volumes between stations 4 + 320 and 4 + 350 and between 4 + 350 and 4 + 400.

$$V = (30/2)(27.0 + 37.8) + (50/2)(37.8 + 55.2) = 3{,}297.9 \text{ m}^3$$

2. For the prismoidal method:

$$V = (L/6)(A_1 + 4A_m + A_2)$$

The middle areas (A_m) are calculated from the averages of the base, top, and height dimensions of the trapezoids.

The middle area between stations 4 + 320 and 4 + 350 is $[(6.0 + 13.5)/2] \times 3.3 = 32.2$ m^2. The middle area between stations 4 + 350 and 4 + 400 is $[(6.0 + 16.5)/2] \times 4.1 = 46.1$ m^2.

The total volume is the sum of the volumes between stations 4 + 320 and 4 + 350 and between 4 + 350 and 4 + 400.

$$V = (30/6)[27.0 + 4(32.2) + 37.8] + (50/6)[37.8 + 4(46.1) + 55.2] = 3{,}279.7 \text{ m}^3$$

Comment: The prismoidal volumes are theoretically more precise than the average end area volumes. The average end area method tends to overestimate volumes, as in this case.

REFERENCES

American Association of State Highway and Transportation Officials (AASHTO). *A Policy on Geometric Design of Highways and Streets.* Washington, DC, 1994.

Garber, N. J. and Hoel, L. A. *Traffic and Highway Engineering.* West Publishing Co., St. Paul, MN, 1988.

Hickerson, T. F. *Route Location and Design,* 5th ed. McGraw-Hill, New York, 1967.

Institute of Transportation Engineers (ITE). *Traffic Engineering Handbook,* 4th ed. Washington, DC, 1992.

Newnan, D. G., ed. *Civil Engineering License Review*, 12th ed. Engineering Press, Austin, TX, 1995.

Transportation Research Board (TRB). *Highway Capacity Manual,* Special Report 209, 3rd ed. TRB, Washington, DC, 1994.

PROBLEMS

6.1   For a simple horizontal curve with $\Delta = 12$ degrees, $R = 400$ m, and PI at Station $0 + 241.782$, the stationing of the PC and PT, respectively, would be most nearly

a. $0 + 197.7$, $0 + 283.8$
b. $0 + 197.7$, $0 + 283.5$
c. $0 + 158.0$, $0 + 283.8$
d. $0 + 195.6$, $0 + 279.4$

6.2   The minimum radius (m) for a simple horizontal curve with a design speed of 110 km/hr is most nearly (assume $e = 0.08$ and $f = 0.14$)

a.   425
b.   395
c.   550
d.   435

6.3   A vehicle hits a bridge abutment at a speed estimated by investigators as 25 km/hr. Skid marks of 30 m on the pavement (coefficient of friction, $f = 0.35$) followed by skid marks of 60 m on the gravel shoulder ($f = 0.50$) approaching the abutment are observed at the accident site. The grade is level. The initial speed (km/hr) of the vehicle was at least

a.   105
b.   90
c.   110
d.   95

For Questions 6.4 and 6.5, assume a $+ 3.9\%$ grade intersects a $+1.1\%$ grade at station $0 + 625.0$ and elevation 305.0 m.

6.4   The minimum length (m) of the vertical curve for a design speed of 80 km/hr is most nearly (assume perception-reaction time $= 2.5$ s, $f = 0.30$, and $S < L$)

a.   85
b.   130
c.   55
d.   110

6.5   The elevation (m) of the middle point of the curve is most nearly

a.   305.00
b.   305.45
c.   304.45
d.   304.55

SOLUTIONS

6.1 b.   Calculate the tangent length:

$$T = R \tan (\Delta/2) = 400 \tan (12/2) = 42.042 \text{ m}$$

Calculate the length of the curve:

$$L = (R\Delta)/57.2958 = 83.776 \text{ m}$$
$$\text{PC sta} = \text{PI sta} - T = 0 + 241.782 - 0 + 042.042 = 0 + 199.740$$
$$\text{PT sta} = \text{PC sta} + L = 0 + 199.740 + 0 + 083.776 = 0 + 283.516$$

6.2 d. $R = V^2/[127(e + f)] = (110)^2/[127(0.08 + 0.14)] = 433 \text{ m}$

6.3 a. The only known speed is the final collision speed of 25 km/hr. Therefore, consider the braking distance on the gravel first.

$$\text{Braking distance} = \left(V_0^2 - V^2\right)/[254 \, (f \pm g)]$$

where V_0 and V represent the initial and final speeds, respectively.

$$\text{Braking distance (gravel)} = 60 \text{ m} = \left(V_0^2 - 25^2\right)/(254)(0.5)$$

Solving for V_0, $V_0 = 90.8$ km/hr = speed at the beginning of the gravel and at the end of the pavement skid. Therefore, for the pavement skid,

$$\text{Braking distance (pavement)} = 30 \text{ m} = \left(V_0^2 - 90.8^2\right)(254)(0.35)$$
$$\text{Solving for } V_0, \; V_0 = 104.5 \text{ km/hr}$$

Comment: Note that this solution does not account for any speed reduction prior to the beginning of the skid marks on the pavement. Therefore, we conclude that the vehicle speed was *at least* 104.5 km/hr.

6.4 b. The required stopping sight distance is

$$S = 0.278 \, Vt + V^2/[254(f \pm g)]$$
$$= 0.278(80)(2.5) + (80)^2/[254(0.30 + 0.011)] = 136.62 \text{ m}$$

The minimum length of crest vertical curve to satisfy the required stopping sight distance is

$$L = (|A|S^2)/404 = [|(1.1 - 3.9)|(136.62)^2]/404 = 129.36 \text{ m}$$

6.5 d. The midpoint offset is

$$E = (AL)/8 = [(0.011 - 0.039)(129.36)]/8 = -0.45 \text{ m}$$

Therefore,

Elevation of the midpoint of the curve
= elevation of the PVI (305 m) − 0.45 m = 304.55 m

Note: The elevation of the midpoint could also be found using the basic properties of the parabola. The offset at the midpoint of the curve ($x = L/2$) is

$$y = ax^2 = [(g_2 - g_1)/2L](L/2)^2 = [(0.011 - 0.039)/2(129.36)]/(129.36/2)^2$$
$$= -0.45 \text{ m}$$

CHAPTER 7

Water Purification and Treatment

Kenneth J. Williamson

WATER DISTRIBUTION

Water distribution systems involve a water source, a transmission line, a distribution network, pumping, and storage.

Water Source

The importance of the water source is primarily related to water elevation and water quality. The water elevation will determine the pumping required to maintain adequate flows in the network.

Water quality is largely determined by the source. Typical water sources include reservoirs, lakes, rivers, and underground aquifers. Surface water sources typically have low levels of dissolved solids but high levels of suspended solids. As a result, treatment of such sources requires coagulation, followed by sedimentation and filtration, to remove suspended solids. Groundwater sources are low in suspended solids but may be high in dissolved solids and reduced compounds. Groundwater, if high in dissolved solids, requires chemical precipitation or ion exchange for removal of dissolved ions.

Transmission Line

A transmission line is required to convey water from the sources and storage to the distribution system. Arterial main lines supply water to the various loops in the distribution system. The main lines are arranged in loops or in parallel to allow for repairs. Such lines are designed using the Hazen-Williams formula for single pipes (Eq. 7.1) and the equivalent-pipe method for single loops.

$$v = kCr^{0.63}s^{0.54}$$ **(7.1)**

where

v = pipe velocity (m/s)
k = constant (2.79)
C = roughness coefficient (100 for cast iron pipe)
r = hydraulic radius (m)
s = slope of hydraulic grade line (m/m)

The equivalent-pipe method involves two procedures: conversion of pipes of unequal diameters in series into a single length of pipe, and conversion of pipes in parallel flow from the same into a single pipe with a single diameter. The headloss is maintained through each conversion. Multiple loops are designed using Hardy Cross methodology.

The Distribution Network

The distribution network comprises the arterial mains, distribution mains, and smaller distribution piping. The controlling design variable for water distribution networks is the pressure under maximum flow, with a minimum value of 140 to 280 kPa and a maximum value of 690 kPa.

Flow for the water distribution network is calculated as the sum of domestic, irrigation, industrial and commercial, and fire requirements. The various flows for a given community are typically obtained from historical data of water use. Without such data, estimates can be made from average values listed in Table 7.1.

Table 7.1 Water flow rates

Flow	Method of Estimation
Average domestic flow	Population served, 0.4 m^3/cap-d, 3,000 to 10,000 persons/km^2
Irrigation flow	Maximum of 0.75 times average daily flow for arid climates
Industrial/commercial flow	Computed from known industries and area of commercial districts

Water Storage

Water storage is required to maintain pressure in the system, to minimize pumping costs, and meet emergency demands. Typically the storage is placed so that the load center or distribution network is between the water source and the storage, as shown in Fig. 7.1.

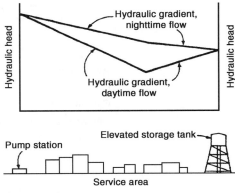

Figure 7.1

Storage requirements are calculated based upon variations in hourly flow and fire requirements. If the pumping capacity is set at some value less than the maximum hourly flow, then all flow requirements above that value will have to be provided by storage.

Pumping Requirements

The pumping system needs adequate capacity for design flows, taking into account the supply from storage. The required head must be adequate to maintain 140 kPa at the load center during maximum flows.

WATER QUALITY

The design of water treatment facilities requires calculations of concentrations and masses of various constituents in water and chemical additions. Common elements and ions of various constituents are listed in Table 7.2, and common water treatment chemicals are listed in Table 7.3. The equivalent weight is equal to the molecular weight divided by the absolute value of the valence.

Ion Balances

Electroneutrality requires that water have an equal number of equivalents of cations and anions. Often, bar graphs are used to show this relationship, as shown in Fig. 7.2. From such diagrams, concentrations of alkalinity, carbonate (calcium and magnesium associated with alkalinity), and noncarbonate hardness (calcium and magnesium in other forms) can be easily identified.

General Chemistry Concepts

A number of concepts from general chemistry are required for engineering calculations related to water treatment.

Oxidation-Reduction Reactions

Oxidation-reduction (redox) reactions involve the transfer of electron from an electron donor to an electron acceptor. The easiest method to construct balanced redox reactions is through the use of half-reactions. A list of common half-reactions related to water treatment are in Table 7.4.

Table 7.2 Common elements and radicals

Name	Symbol	Atomic Weight	Valence	Equivalent Weight
Aluminum	Al	27.0	+3	9.0
Calcium	Ca	40.1	+2	20.0
Carbon	C	12.0	−4	
Chlorine	Cl	35.5	−1	35.5
Fluorine	F	19.0	−1	19.0
Hydrogen	H	1.0	+1	1.0
Iodine	I	126.9	−1	126.9
Iron	Fe	55.8	+2	27.9
			+3	
Magnesium	Mg	24.3	+2	12.15
			+4	
			+7	
Nitrogen	N	14.0	−3	
			+5	
Oxygen	O	6.0	−2	8.0
Potassium	K	39.1	+1	39.1
Sodium	Na	23.0	+1	23.0
Ammonium	NH_4^+	18	+1	18.0
Hydroxyl	OH^-	17.0	−1	17.0
Bicarbonate	HCO_3^-	61.0	−1	61.0
Carbonate	CO_3^{2-}	60.0	−2	30.0
Nitrate	NO_3^-	46.0	−1	46.0
Hypochlorite	OCl^-	51.5	−1	51.5

Henry's Law

Henry's Law states that the weight of any dissolved gas is proportional to the pressure of the gas.

$$C_{equil} = \alpha \, p_{gas} \tag{7.2}$$

where

C_{equil} = equilibrium dissolved gas concentration
P_{gas} = partial pressure of gas above liquid
α = Henry's Law constant

Equilibrium Relationships

For an equilibrium chemical equation expressed as

$$A + B \rightleftharpoons C + D \tag{7.3}$$

the relationship of concentrations at equilibrium can be approximated as

$$K_{eq} = \frac{[C][D]}{[A][B]} \tag{7.4}$$

where

K_{eq} = equilibrium constant
[ ] = molar concentrations

Table 7.3   Common inorganic chemicals for water treatment

Name	Formula	Usage	Molecular Weight	Equivalent Weight
Activated carbon	C	Taste and odor	12.0	
Aluminum sulfate	$Al_2(SO_4)_3 \cdot 14.3H_2O$	Coagulation	600	100
Ammonia	NH_3	Chloramines, disinf.	17.0	
Ammonium fluosilicate	$(NH_4)_2SiF_6$	Fluoridation	178	
Calcium carbonate	$CaCO_3$	Corrosion control	132	66.1
Calcium fluoride	CaF_2	Fluoridation	78.1	
Calcium hydroxide	$Ca(OH)_2$	Softening	74.1	37.0
Calcium hypochlorite	$Ca(ClO)_2 \cdot 2H_2O$	Disinfection	179	
Calcium oxide	CaO	Softening	56.1	28.0
Carbon dioxide	CO_2	Recarbonation	44.0	22.0
Chlorine	Cl_2	Disinfection	71.0	
Chlorine dioxide	ClO_2	Taste and odor	67.0	
Ferric chloride	$FeCl_3$	Coagulation	162	54.1
Ferric hydroxide	$Fe(OH)_3$		107	35.6
Fluorosilicic acid	H_2SiF_6	Fluoridation	144	16.0
Oxygen	O_2	Aeration	32.0	
Sodium bicarbonate	$NaHCO_3$	pH adjustment	84.0	84.0
Sodium carbonate	Na_2HCO_3	Softening	106	53.0
Sodium hydroxide	$NaOH$	pH adjustment	40.0	40.0
Sodium hypochlorite	$NaClO$	Disinfection	74.4	
Sodium fluosilicate	Na_2SiF_6	Fluoridation	188	

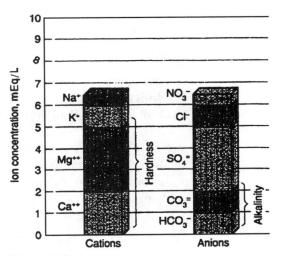

Figure 7.2

Some commonly used equilibrium constants are listed in Table 7.5.

A common equilibrium relationship is the disassociation of water, which is given as

$$H_2O \rightleftharpoons H^+ + OH^- \tag{7.5}$$

which has a K_{eq} value of 10^{-7}. The hydrogen ion concentration is typically represented by the pH value, or the negative log of the hydrogen ion concentration:

$$pH = -\log[H^+] \tag{7.6}$$

Table 7.4 Common half-reactions for water treatment

Reduced Element	Half-Reaction
Cl	$\frac{1}{2}Cl_2 + e^- \rightarrow Cl^-$
Cl	$\frac{1}{2}ClO^- + H^+ + e^- \rightarrow \frac{1}{2}Cl^- + \frac{1}{2}H_2O$
Cl	$\frac{1}{8}ClO_4^- + H^+ + e^- \rightarrow \frac{1}{8}Cl^- + \frac{1}{2}H_2O$
Fe	$\frac{1}{2}Fe^{2+} + e^- \rightarrow \frac{1}{2}Fe$
Fe	$Fe^{3+} + e^- \rightarrow Fe^{2+}$
Fe	$\frac{1}{3}Fe^{3+} + e^- \rightarrow \frac{1}{3}Fe$
I	$\frac{1}{2}I_2 + e^- \rightarrow I^-$
N	$\frac{1}{8}NO_3^- + 5/4H^+ + e^- \rightarrow \frac{1}{8}NH_4^+ + \frac{3}{8}H_2O$
N	$\frac{1}{5}NO_3^- + \frac{6}{5}H^+ + e^- \rightarrow \frac{1}{10}N_2 + \frac{3}{5}H_2O$
O	$\frac{1}{4}O_2 + H^+ + e^- \rightarrow \frac{1}{2}H_2O$

Table 7.5 Common equilibrium constants

Equation	K_{eq}
$H_2CO_3 \rightleftharpoons H^+ + HCO_3^-$	$10^{-6.4}$
$HCO_3^- \rightleftharpoons H^+ + CO_3^{2-}$	$10^{-10.3}$
$NH_3 + H_2O \rightleftharpoons NH_4^+ + OH^-$	$10^{-4.7}$
$CaOH^+ \rightleftharpoons Ca^{2+} + OH^-$	$10^{-1.5}$
$MgOH^+ \rightleftharpoons Mg^{2+} + OH^-$	$10^{-2.6}$
$HOCl \rightleftharpoons H^+ + OCl^-$	$10^{-7.5}$

Alkalinity

Alkalinity is a measure of the ability of water to consume a strong acid. In the alkalinity test, water is titrated with a strong acid to pH 6.4 and to pH 4.5. The amount of acid that is used to reach the first pH endpoint is termed the **carbonate alkalinity,** and the amount of acid needed to reach the second end point is the **total alkalinity**. Alkalinity is expressed as calcium carbonate.

Alkalinity is the sum of bicarbonate, carbonate, and hydroxide concentrations minus hydrogen ion concentration. For most sources to be used for drinking water, alkalinity can be approximated as carbonate and bicarbonate. Carbonate alkalinity is the predominant species when the pH is above 10.8, and bicarbonate is the predominant species when the pH is between 6.9 and 9.8.

Solubility Relationships

The equilibrium between a compound in its solid crystalline state and its ionic form in solution, where

$$X_aY_b = aX^{b+} + bY^{a-} \tag{7.7}$$

is given by

$$K_{sp} = [X^{b+}]^a[Y^{a-}]^b \tag{7.8}$$

Table 7.6 Common solubility products

Compound	K_{sp}
Magnesium carbonate	4×10^{-5}
Magnesium hydroxide	9×10^{-12}
Calcium carbonate	5×10^{-9}
Calcium hydroxide	8×10^{-6}
Aluminum hydroxide	1×10^{-32}
Ferric hydroxide	6×10^{-38}
Ferrous hydroxide	5×10^{-15}
Calcium fluoride	3×10^{-11}

where K_{sp} = solubility product. Solubility products for common precipitation reactions in water treatment are given in Table 7.6.

Dissolved Oxygen Relationships

The dissolved oxygen concentration in a stream is determined by the rate of oxygen consumption, commonly caused by the discharge of oxygen-demanding substances, and the rate of reoxygenation from the atmosphere.

Biochemical Oxygen Demand

The concentration of oxygen-demanding substances in a river are commonly expressed as biochemical oxygen demand or BOD as described in Chapter 5. BOD is measured as the oxygen that is removed in a 20°C, five-day test and is expressed as BOD_5. This value has to be converted to an ultimate BOD or L value at the temperature of interest as

$$L = \frac{BOD_5}{1 - e^{-5k_{20}}} \tag{7.13}$$

where k_{20} = BOD decay constant at 20°C. Typical k_{20} values for organic wastes range from about 0.20 to 0.30.

The rate of oxygen consumption by BOD is related to the ultimate BOD concentration as

$$r_{BOD} = k_T L \tag{7.14}$$

Oxygen Deficit

The dissolved oxygen in water is determined by Henry's Law, based on a 20 percent partial pressure in the atmosphere. Saturation values as a function of temperature can be extrapolated from known values of 14.6, 11.3, 9.1, and 7.5 mg/L at 0, 10, 20, and 30°C, respectively.

At a given temperature the dissolved oxygen concentration can be expressed also as a deficit or

$$D = (C_{sat} - C) \tag{7.15}$$

where

C_{sat}, C = saturation and actual dissolved oxygen concentration (mg/L)
D = dissolved oxygen deficit at given temperature

Mixing

Assuming complete mixing in a river, the concentration of dissolved oxygen and ultimate BOD in a receiving stream are related as

$$C_0 = \frac{Q_r C_r + Q_w C_w}{Q_r + Q_w} \tag{7.16}$$

where

C_r, C_w = concentration of constituents in river and waste, respectively
Q_r, Q_w = flow rate of river and waste, respectively

Reoxygenation

Reoxygenation occurs by diffusion of oxygen from the atmosphere to the river. The rate of reoxygenation or reaeration is given as

$$r_{\text{reoxy}} = k_2 D \tag{7.17}$$

where k_2 = reaeration coefficient (1/d).

Oxygen Sag Model

The dissolved oxygen concentration in a stream after the addition of oxygen-demanding wastes can be expressed in differential form as

$$\frac{dD}{dt_r} = r_{\text{BOD}} - r_{\text{reoxy}} \tag{7.18}$$

where t_r = travel time in the river from the point of waste addition (d).

The resulting dissolved oxygen profile is shown in Fig. 7.3. After addition of the waste, the dissolved oxygen decreases to a maximum deficit (D_c) or minimum dissolved oxygen level at a distance x_c or travel t_c. Past this point, the deficit decreases as the river recovers.

Equation (7.18) can be integrated to

$$D = \frac{kL_0}{k_2 - k}\left(e^{-kt_r} - e^{-k_2 t_r}\right) + D_0 e^{-k_2 t_r} \tag{7.19}$$

where D_0, L_0 = oxygen deficit and ultimate BOD concentration at point of waste discharge in the river, respectively.

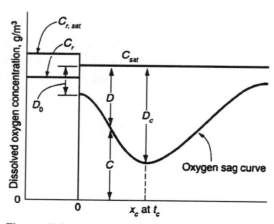

Figure 7.3

WATER TREATMENT

Water treatment is used to alter the quality of water to make it chemically and bacteriologically safe for human consumption. Common sources are groundwater and surface waters. Groundwater treatment may involve removal of pathogens, removal of iron and manganese, and removal of hardness (Fig. 7.4). Surface water treatment typically involves simultaneous removal of pathogens, suspended solids, and taste- and odor-causing compounds (Fig. 7.5). All of these treatment processes are composed of unit processes.

Sedimentation

After flocculation, the larger particles are removed by sedimentation. Settling is also used in water treatment after oxidation of iron or manganese. Flocculated particles have densities of about 1400 to 2000 kg/m.

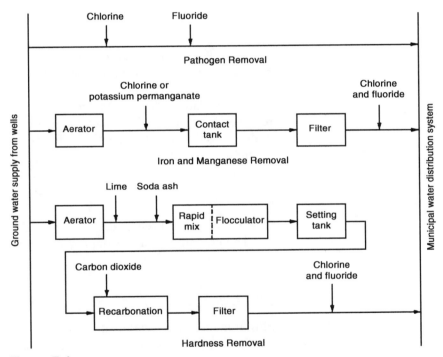

Figure 7.4

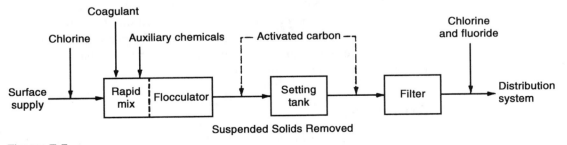

Figure 7.5

Settling velocities of particles are determined by Stokes' law for Reynolds numbers less than 0.3:

$$v_s = \frac{g(\rho_p - \rho_w)d_p^2}{18\mu} \tag{7.20}$$

where

v_s = terminal settling velocity (m/s)
g = gravitational constant (9.8 m/s^2)
d_p = particle diameter (m)
μ = water viscosity (0.001 kg/m-s)

For particles coming into a sedimentation basin, the particles enter at all depths. A critical settling velocity related to the sedimentation basin can be calculated as

$$v_{sc} = \frac{Q}{A_s} \tag{7.21}$$

where

Q = flow rate into sedimentation basin (m^3/s)
A_s = surface area of sedimentation basin (m^2)

In standard sedimentation theory, if the settling rate of the particles to be removed is greater than v_{sc}, then the particles will be 100% removed. For particles with v_s less than v_{sc}, the particles will be only partially removed, with the removal given as

$$R = \frac{v_s}{v_{sc}} \tag{7.22}$$

where R = decimal removal.

The critical settling velocities used for design are typically expressed as an overflow rate.

Filtration

Filtration in water treatment is commonly accomplished with granular filters. Media for such filters may include sand, charcoal, and garnet; the filters can involve one, two, or several different filter media. The water is applied to the top of the filter and is collected through underdrains.

Filtration is a complex process involving entrapment, straining, and absorption. As the filtration process proceeds, the headloss associated with the water flow through the porous media increases as the filtered material accumulates in the pores. In addition, the number of particles passing through the filter increases, with a subsequent increase in effluent turbidity. When either the allowable headloss or allowable effluent quality is exceeded, then the filter has to be backwashed to remove the accumulated material.

Softening

The removal of hardness is accomplished using line/soda ash softening or ion exchange.

Lime/Soda Ash Softening

Lime/soda ash softening occurs by the addition of calcium hydroxide and sodium bicarbonate to form a chemical precipitate, which is removed by sedimentation and filtration.

For waters containing excess, the calcium hydroxide reacts with carbon dioxide and carbonate hardness as

$$CO_2 + Ca(OH)_2 \rightarrow CaCO_3 + H_2O$$
$$Ca(HCO_3)_2 + Ca(OH)_2 \rightarrow CaCO_3 + 2H_2O$$
$$Mg(HCO_3)_2 + 2Ca(OH)_2 \rightarrow 2CaCO_3 + Mg(OH)_2 + 2H_2O$$

The sodium bicarbonate and calcium hydroxide react with the noncarbonate hardness as

$$Ca^{2+} + Na_2CO_3 \rightarrow CaCO_3 + H_2O$$
$$Mg^{2+} + Ca(OH)_2 \rightarrow Mg(OH)_2 + Ca^{2+}$$

Recarbonation is required after treatment to remove the excess lime, magnesium hydroxide, and carbonate, reducing the pH to about 8.5 to 9.5.

Based upon the foregoing equations, the requirements of lime (L) and soda ash (SA) in mEq/L are

$$L = CO_2 + HCO_3^- + Mg^{2+} \tag{7.23}$$

$$SA = Ca^{2+} + Mg^{2+} - Alk \tag{7.24}$$

The lime requirement needs to be increased about 1 Eq/m^3 to raise the pH to allow precipitation of the magnesium hydroxide.

Ion Exchange

Ion exchange processes are used to remove ions of calcium, magnesium, iron, and ammonium. The exchange material is a solid that has functional groups that replace ions in solution for ions on the exchange material. Ion exchange materials can be made to remove cations or anions. Most materials are regenerated by flushing with solutions with high concentrations of either H^+ or Na^+.

Ion exchange process is controlled by the exchange capacity and the selectivity. The **exchange capacity** expresses the equivalents of cations or anions that can be exchanged per unit mass. The **selectivity** for ion B as compared to ion A is expressed as

$$K_{B/A} = \frac{\chi_{R-A.}[A^+]}{\chi_{R-B}[B^+]} \tag{7.25}$$

where

$K_{B/A}$ = selectivity coefficient for replacing A on resin with B;

χ_{R-A}, χ_{R-B} = mole fractions of A and B for the absorbed species.

Chlorination

Chlorination is the most common form of disinfection for treatment of water. Chlorine is a strong oxidation agent, which destroys organisms by chemical attack. Chlorine is usually added in the form of chlorine gas at a level of 2 to 5 mg/L. Other common forms are chlorine dioxide, sodium hypochlorite, and calcium hypochlorite.

Chlorine in water undergoes an equilibrium reaction to form HOCl and OCl⁻. The hypochlorite is a weak acid as listed in Table 7.5. HOCl is a more effective disinfectant than OCl⁻; as such, the effectiveness is strongly dependent upon pH.

The effectiveness of chlorination depends upon the time of contact, the chlorine concentration, and the concentration of organisms. The effect of time is modeled as

$$\frac{N_t}{N_0} = e^{-kt^m} \tag{7.26}$$

where

N_t, N_0 = number of organisms at time t and time zero
k = decay coefficient (1/d)
t = time (d)
m = empirical coefficient (usually 1)

Fluoridation

Fluoridation of water supplies has been shown to reduce dental caries dramatically. Optimum concentrations are about 1 mg F/L. The most commonly used compounds are sodium fluoride, sodium fluosilicate, and fluorosilicic acid. Fluoride is usually added after coagulation or lime/soda ash softening, because high calcium concentration can result in the fluorides precipitating.

Activated Carbon

Activated carbon is added in water treatment to adsorb taste- and odor-causing compounds and to remove color. The process usually involves the direct addition of powdered activated carbon, with removal in the sedimentation and filtration units. The required dosage is usually determined empirically using laboratory tests.

Sludge Treatment

The sources of sludge in water treatment include sand, silt, chemical sludges, and backwash solids. These sludge are typically concentrated in settling basins or lagoons. The sludges can be further concentrated by drying on sand drying beds or by centrifugation.

PROBLEMS

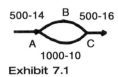

500-14 B 500-16

A   C

1000-10

Exhibit 7.1

7.1 Find a single pipe to replace the pipe loop shown in Exhibit 7.1.

7.2 How many grams of oxygen are required to burn 1 gram of methane?

$$CH_4 + O_2 \rightarrow CO_2 + H_2O$$

7.3 A liter of water is at equilibrium with an atmosphere containing a partial pressure of 0.1 atm of CO_2. How many grams are dissolved in the water? ($\alpha = 2.0$ g/L-atm)

7.4 How much HOCl is present in a solution containing 0.1M chlorine at pH 8?

7.5 How many grams of fluoride would be present in a solution saturated with calcium fluoride?

7.6 A 5-d BOD and ultimate BOD are measured at 180 mg/L and 200 mg/L, respectively. What is the decay coefficient?

7.7 A wastewater with a dissolved oxygen concentration of 1 mg/L is discharged to a river. The river is at 20°C and saturated with dissolved oxygen. If the flows are 2.8×10^{-2} m/s and 2.8 m/s for the waste water and the river, respectively, what is the oxygen deficit after mixing?

7.8 A water treatment plant is required to produce an average treated water flow 14.4×10^4 m^3/d from a water source of the following characteristics:

Chlorines	92 mg/L
Potassium	31 mg/L
Sodium	14 mg/L
Sulfates	134 mg/L
Calcium	94 mg/L
Magnesium	28 mg/L
Alkalinity	135 mg/L as $CaCO_3$
pH	7.8
Temperature	21°C
Total solids	720 mg/L

Determine an ion balance for the water.

7.9 A water contains silt particles with a uniform diameter of 0.02 mm and a specific gravity of 2.6. What removal is expected in a clarifier with an overflow rate of 12 m/d (300 gal/ft^2-d)?

SOLUTIONS

7.1   Assume a flow in A-B-C of 4 cfs. Use the nomograph in Exhibit 7.1a.

$$h_L \text{ in AB} = 5.8'/1000'$$
$$= 2.9 \text{ ft}$$

$$h_L \text{ in BC} = 3.0'/1000'$$
$$= 1.5'$$

$$\text{Total } h_L = 4.5'$$
$$= 4.5'/1000'$$

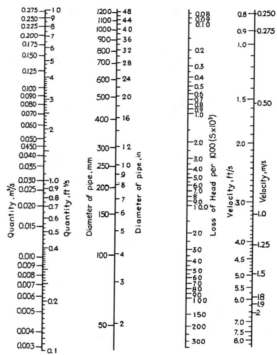

Exhibit 7.1a   Flow in old cast iron pipes. (Hazen-Williams $C = 100$)

Use $\phi = 14.5"$, L = 1000' for a h_L of 4.5' for A-C. then $Q = 1.4$ cfs.

$$Q_{\text{Total}} = 4 + 1.4 = 5.4 \text{ cfs}$$
$$\text{Equivalent diameter for loop} = 16.2"$$
$$\text{Equivalent length for loop} = 1000'$$

7.2   Balance the equation as

$$CH_4 + 2O_2 \rightarrow CO_2 + 2H_2O$$
$$gO_2 = 1 \text{ g } CH_4 \times \frac{1 \text{ mol } CH_4}{16 \text{ g } CH_4} \times \frac{2 \text{ moles } O_2}{1 \text{ mole } CH_4} \times \frac{32 \text{ g } O_2}{1 \text{ mole } O_2}$$
$$= 4 \text{ g } O_2$$

7.3

$$C_{\text{equil}} = \alpha\, P \text{ gas}$$
$$= 2.0\,\text{g}/1 - \text{atm} \times 0.1\,\text{atm}$$
$$= 2.0\,\text{g}/\text{liter}$$

$$M_{\text{CO}_2} = C \times V$$
$$= 0.2\,\text{g}/1 \times 1\,\text{liter}$$
$$= 0.2\,\text{g}$$

7.4

$$[\text{H}^+] = 10^{-\text{pH}} = 10^{-8}$$
$$\text{HOCl} \Leftrightarrow \text{H}^+ + \text{OCl}^-$$
$$\frac{[\text{H}^+][\text{OCl}^-]}{[\text{HOCl}]} = 10^{-7.5}$$
$$\frac{[\text{H}^+](0.1 - \text{HOCl})}{[\text{HOCl}]} = 10^{-7.5}$$
$$\frac{(0.1 - [\text{HOCl}])}{[\text{HOCl}]} = 3.16$$
$$[\text{HOCl}] = 0.024\ \text{M}$$

7.5

$$\text{CaF}_2 \rightarrow \text{Ca}^{2+} + 2\text{F}^-$$
$$[\text{Ca}^{2+}][\text{F}^-]^2 = 3 \times 10^{-11}$$
$$\left[\frac{1}{2}\text{F}^-\right][\text{F}^-]^2 = 3 \times 10^{-11}$$
$$[\text{F}^-] = [2(2 \times 10^{-11})]^{1/3}$$
$$= 4.2 \times 10^{-4}\ \text{M}$$
$$= 8.0 \times 10^{-3}\ \text{g/liter}$$

7.6

$$1 - e^{-k(5d)} = \text{BOD}_5/\text{L}$$
$$= 180/300 = 0.6$$
$$k = 0.18/\text{d}$$

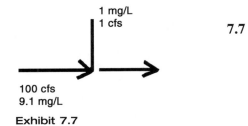

1 mg/L
1 cfs

100 cfs
9.1 mg/L

Exhibit 7.7

7.7

$$C = \frac{(100\ \text{cfs})(9.1\ \text{mg/L}) + (1\ \text{cfs})(1\ \text{mg/L})}{101\ \text{cfs}}$$
$$= 9.0\ \text{mg/L}$$
$$D = C_{\text{sat}} - C$$
$$= 0.1\ \text{mg/L}$$

7.8   The milliequivalents per liter for the ions are:

Ion	Conc (mg/L)	Equiv. wt.	Conc (mEq/L)
Na^+	14	23	0.61
K^+	31	39	0.79
Mg^{2+}	28	12.2	2.30
Ca^{2+}	94	20	4.7
		Total	8.4
SO_4^{2-}	134	48	2.79
Cl^-	92	35.5	2.59
$HCO_3^- + CO_3^{2-}$	135	50	2.70
		Total	8.1

When the concentration of the alkalinity (expressed as $CaCO_3$) is converted to mEq/L, it will be equal to the sum of the concentrations of bicarbonate and carbonate.

7.9

$$v_s = \frac{g(\rho_p - \rho_w)d_p^2}{18\mu}$$

$$= \frac{(9.8 \text{ m/s}^2)(1600 \text{ kg/m}^3)(2 \times 10^{-5} \text{ m})^2}{(18)(1 \times 10^{-3} \text{ N-s/m}^2)}$$

$$= 3.5 \times 10^{-5} \text{ m/s}$$

$$v_{sc} = \frac{300 \text{ gal}}{\text{ft}^2\text{-d}} \times \frac{0.0017 \text{ m/hr}}{1 \text{ gal/ft}^2\text{-d}} \times \frac{1 \text{ hr}}{3600 \text{ s}}$$

$$v_{sc} = 12 \text{ m/d} \times \frac{d}{86,400 \text{ s}} = 1.4 \times 10^{-4} \text{ m/s}$$

$$= 1.4 \times 10^{-4} \text{ m/s}$$

$$k = \frac{3.5 \times 10^{-5} \text{ m/s}}{14 \times 10^{-5} \text{ m/s}} = 0.25 \text{ or } 25\%$$

Computers and Numerical Methods

Lincoln D. Jones

INTRODUCTION

The *Fundamentals of Engineering Exam* contains seven questions concerning computers in the morning session and three questions in the afternoon session. These questions cover the topics of operating systems, networks, interfaces, spreadsheets, flow charting, and data transmission. Each of the branch-specific afternoon exams contains three questions on numerical methods related to that branch.

You should review the glossary of important computer-related keywords that is presented at the end of this chapter to ensure that you have a basic understanding of this broad general topic. Should you find any term unfamiliar to you, a review of that topic would be warranted. The current exam does not include a programming language such as FORTRAN or BASIC, but use of applications such as spreadsheets is included. You should be familiar with one of the popular spreadsheet programs such as Excel, Quattro Pro, or 1-2-3.

NUMBER SYSTEMS

The number system most familiar to everyone is the decimal system based on the ten symbols 0 through 9. This base-10 system requires ten different digits to create the representation of numbers.

A far simpler system is the binary number system. This base-2 system uses only the characters 0 and 1 to represent any number. A binary representation 110, for example, corresponds to the number

$$1 \times 2^2 + 1 \times 2^1 + 1 \times 2^0 = 4 + 2 + 0 = 6$$

Similarly, the binary number 1010 would be

$$1 \times 2^3 + 0 \times 2^2 + 1 \times 2^1 + 0 \times 2^0 = 8 + 0 + 2 + 0 = 10$$

The digital computer is based on the binary system of on/off, yes/no, or 1/0. For this reason it may at times be necessary to convert a decimal number (such as 12) to a binary number (for 12, it would be 1100). One binary digit is known as a **bit**, the smallest possible unit of information. Eight bits make up a **byte** (B).

Example **8.1**

The binary number 1110 corresponds to what decimal (base 10) number?

Solution

$$1 \times 10^3 + 1 \times 10^2 + 1 \times 10^1 + 0 \times 10^0 = 8 + 4 + 2 = 14$$

DATA STORAGE

Memory chips include random-access memory (RAM), read-only memory (ROM), and programmable read-only memory (PROM).

Diskettes usage have been the primary way to move data between personal computers. Initially diskettes were 8-inch or $5\frac{1}{4}$-inch "floppies" (360 KB to 2 MB of data), which could be damaged easily. In the 1980s they were superseded by $3\frac{1}{2}$-inch diskettes (720 KB to 1.44 MB of data) in a hard plastic housing.

Hard drives (currently about 10–200 GB of data) are units with the hard disks permanently sealed in a module permanently installed in a computer. Other storage devices are removable hard disk cartridges; optical disks; read-only media (ROM); write-once, read-many media (WORM); and compact disk read-only memory (CD-ROM). The CD-ROM has become very common for carrying large files.

DATA TRANSMISSION

A computer produces digital signals (represented as on–off or 0–1), but these signals often must be transmitted on voice transmission (analog) systems. The conversion of digital signals to analog signals is called **modulation**. The subsequent conversion back to digital from analog is called **demodulation**. The device doing this conversion is called a **modem** (short for modulation-demodulation). Early modems had speeds of 300–1200 bits per second, but modern modems operate at 14,400–28,800 bits per second.

Data transmission requires that the equipment at both the sending and receiving locations must be able to "talk" to each other. Two methods commonly used are asynchronous and synchronous transmission. In **asynchronous** transmission a start signal is sent at the beginning and a stop signal at the end of each character. In **synchronous** transmission, on the other hand, a bit pattern is transmitted at the beginning of the message to synchronize the internal clocks of the sending and receiving devices.

PROGRAMMING

Computer programming may be thought of as a four-step process:

1. Defining the problem

2. Planning the solution

3. Preparing the program

4. Testing and documenting the program

Once the problem has been carefully defined, the basic programming work of planning the computer solution and preparing the detailed program can proceed. In this section the discussion will be limited to two ways to plan a computer program to solve a problem:

- Algorithmic flowcharts

- Pseudocode

Algorithmic Flowcharts

An algorithmic flowchart is a pictorial representation of the step-by-step solution of a problem using standard symbols. Some of the commonly used symbols are shown in Fig. 8.1.

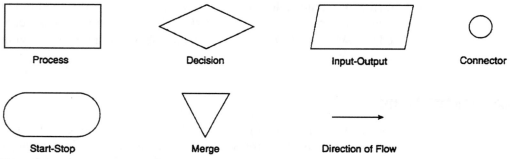

Process   Decision   Input-Output   Connector

Start-Stop   Merge   Direction of Flow

Figure 8.1 Flowchart symbols

Example 8.2

A present sum of money (*P*) at an annual interest rate (*I*), if kept in a bank for *N* years, would amount to a future sum (*F*) at the end of that time according to the equation $F = P(1 + I)^N$. Prepare a flowchart for *P* = $100, *I* = 0.07, *N* = 5 years, and compute and output the values of *F* for all values of *N* from 1 to 5.

Solution

Exhibit 1 is a flowchart for this situation.

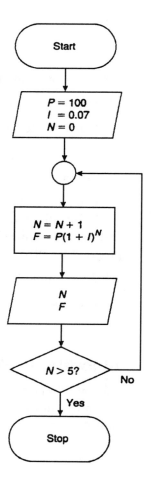

Start

$P = 100$
$I = 0.07$
$N = 0$

$N = N + 1$
$F = P(1 + I)^N$

N
F

$N > 5?$ No

Yes

Stop

Exhibit 1

Example **8.3**

Consider the flowchart in Exhibit 2.

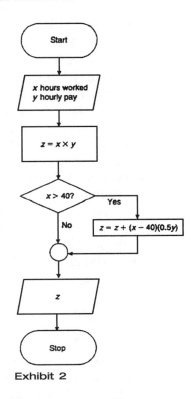

Exhibit 2

The computation does which of the following?

(a) Inputs hours worked and hourly pay and outputs the weekly paycheck for 40 hours or less

(b) Inputs hours worked and hourly pay and outputs the weekly paycheck for all hours worked including over 40 hours at premium pay

Solution

The answer is (b).

Pseudocode

Pseudocode is an English-like language representation of computer programming. It is carefully organized to be more precise than a simple statement, but it may lack the detailed precision of a flowchart or of the computer program itself.

Example **8.4**

Prepare pseudocode for the computer problem described in Example 8.2.

Solution

```
INPUT P, I, and N = 0
DOWHILE N < 5
COMPUTE N = N + 1
      F = P (1 + I)ᴺ
OUTPUT   N, F
IF N > 5 THEN ENDO
```

SPREADSHEETS

For today's engineers the ability to create and use spreadsheets is essential. The most popular spreadsheet programs are Microsoft's **Excel**, Corel's **Quattro Pro**, and Lotus **1-2-3**. Each of these programs uses similar construction, methods, operators, and relative references.

Three types of information may be entered into spreadsheets: text, values, and formulas. Text includes labels, headings, and explanatory text. Values are numbers, times, or dates. Formulas combine operators and values in an algebraic expression.

A **cell** is the intercept of a column and a row. Its location is based upon its column-row location, for example B3 would be the intercept of column B and row 3. Column labels are across the spreadsheets and row labels are on the side. To change a cell entry, the cell must be selected using either an address or a pointer.

A group of cells may be called out by using a **range**. Cells A1, A2, A3, and A4 could be called out using the range reference A1:A4 (or A1..A4). Similarly, A2, B2, C2, and D2 would use the range reference A2:D2 (or A2..D2).

In order to call out a block of cells, a range callout might be A2:C4 (or A2..C4) and would reference the following cells:

A2   B2   C2

A3   B3   C3

A4   B4   C4

Formulas may include cell references, operators (such as +, −, *, and /) and functions (such as SUM and AVG). The formula SUM(A2:A6) or SUM(A2..A6) would be evaluated as equal to A2 + A3 + A4 + A5 + A6.

Relational References

Most spreadsheet references are relative to the cell's position. For example, if the content of cell A5 contained B4 then the value of A5 is the value of the cell up one and over one. The relational reference is most frequently used in table tabulations as the following example for an inventory, where cost times quantity equals value and the sum of the values yields the total inventory cost.

Inventory valuation

	A	B	C
1 Item	Cost	Quantity	Value
2 box	5.2	2	10.4
3 tie	3.4	3	10.2
4 shoe	2.4	2	4.8
5 hat	1.0	1	1.0
6 Sum			26.4

In C2 use the formula A2*B2.
In C3 use the formula A3*B3 and so forth.
For the summation use the function SUM. In C6 use SUM(C2:C5)

Instead of typing in each cell's formula, the formula can be copied from the first cell to all of the subsequent cells in many common spreadsheet applications by first highlighting cell C2, then dragging the mouse to include cell C5. The first active cell, C2, would be displayed in the edit window. Type the formula for C2 as A2*B2 and hold the control key down when the enter key is pressed. This copies the relational formula to each of the highlighted cells. Since the call is relational, then in cell C3 the formula is evaluated as A3*B3. In C4 the cell is evaluated as A4*B4. Similarly, edit operations such as copy, paste, and fill operations simplify the duplication of relational formulas from previously filled-in cells.

Arithmetic Order of Operation

Operations in equations use the following sequence: exponentiation, followed by multiplication and division, followed by addition and subtraction. Parentheses in formulas override normal operator order.

Absolute References

Sometimes a reference to a cell must be used that should not change with the location of the reference, such as a reference to a cell containing a data variable. An **absolute reference** can be specified by inserting a dollar sign "$" before the column-row reference. If B2 is the data entry cell, then by using B2 as its reference in another cell, the call will always be evaluated to cell B2, even though it may be copied to another cell. Mixed references can be made by using the dollar sign for only one of the elements of the reference. The reference B$2 is a mixed reference in that the row does not change but the column remains a relational reference.

The power of spreadsheets makes repetitive calculations very easy. In many operations, periods of time are normally new columns. Each element becomes a row item with changes in time becoming the columns.

All spreadsheet programs allow for changes in the appearance of the spreadsheet. Headings, borders, or type fonts are usual customizing tools.

NUMERICAL METHODS

This section includes techniques of finding roots of polynomials using the Routh-Hurwitz criterion and Newton methods, Euler's techniques of numerical integration and the trapezoidal methods, and techniques for the determination of numerical solutions of differential equations.

Root Extraction

Routh-Hurwitz Method (Without Actual Numerical Results):

Root extraction, even for simple roots (i.e., without imaginary parts), can become quite tedious. Before attempting to find roots, one should first ascertain whether they are really needed or whether just knowing the area of location of these roots will suffice. If all that is needed is knowing whether the roots are all in the left half-plane of the variable (such as in the s-plane when using Laplace transforms, which is frequently the case in determining system stability in control systems), then one may use the Routh-Hurwitz criterion. This method is fast and easy even for higher-ordered equations. As an example, consider the following polynomial:

$$P(x) = \prod_{m=1}^{n}(x - a_m) = x^n + a_1 x^{n-1} + a_2 x^{n-2} + \Lambda + a_{n-1} \tag{8.1}$$

Here, finding the roots for $n > 3$ can become quite tedious without a computer; however, if one only needs to know if any of the roots have positive real parts, then use the Routh-Hurwitz method. Here, an array may be made listing the coefficients of every other term starting with the highest power, n, on a line, then a following line may be made listing the coefficients of the terms left out of the first row. Following rows are constructed using Routh-Hurwitz techniques, and after completion of the array, one merely checks to see if all the signs are the same (unless there is a zero coefficient—then something else needs to be done) in the first column; if none, no roots will exist in the right half-plane. This is a simple technique used in control systems (for details, see almost any text dealing with stability of control systems). A short example follows:

$$F(s) = s^3 + 3s^2 + 2s + 10$$
$$= (s + ?)\,(s + ?)\,(s + ?)$$

Array:

s^3	1	2
s^2	3	10
s^1	−4/3	0
s^0	10	0

where the s^1 term is formed as $(3 \times 2 - 10 \times 1)/3 = -4/3$. For details, refer to any text on control systems or numerical methods.

Here, there are two sign changes, one from 3 to −4/3, and one from −4/3 to 10; this means there will be two roots in the right half-plane of the s-plane, which yields an unstable system; this technique represents a great savings in one's time without having to actually factor the equation.

Newton's Method

The use of Newton's method of solving a polynomial and the use of iterative methods can greatly simplify the problem. Newton's method utilizes synthetic division and is based upon the remainder theorem. This synthetic division requires estimating a root at the start, and, of course, the best estimate is the actual root. The root is the correct one when the remainder is zero. (There are several ways of estimating this root, including a slight modification of the Routh-Hurwitz criterion.)

By taking a $P_n(x)$ polynomial [see Eq. (8.1)] and dividing it by an estimated factor $(x - x_1)$, the result is a reduced polynomial of degree $n - 1$, $Q_{n-1}(x)$, plus a constant remainder of b_{n+1}. Thus, another way of describing Eq. (8.1) is

$$\frac{P_n(x)}{(x - x_1)} = \frac{Q_{n-1}(x) + b_{n-1}}{(x - x_1)} \quad \text{or} \quad P_n(x) = (x - x_1)Q_{n-1}(x) + b_{n-1} \qquad \textbf{(8.2)}$$

If one lets $x = x_1$, Eq. (8.2) becomes

$$P_n(x = x_1) = (0)Q_{n+1}(x) + b_{n+1} = b_{n+1}. \qquad \textbf{(8.3)}$$

Equation (8.3) leads directly to the remainder theorem: "The remainder on division by $(x - x_1)$ is the value of the polynomial at $x = x_1$, $P_n(x_1)$." [1]

Newton's method (actually, the Newton-Raphson method) for finding the roots for an nth-order polynomial is an iterative process involving obtaining an estimated value of a root (leading to a simple computer program). The key to the process is getting the first estimate of a possible root; without getting too involved, recall that the coefficient of x^{n-1} represents the sum of all of the roots, and the last term represents the product of all n roots. Then the first estimate can be "guessed" within a reasonable magnitude. After a first root is chosen, find the rate of change of the polynomial at the chosen value of the root to get the next closer value of the root, x_{n+1}. Thus the new root estimate is based on the last value chosen,

$$x_{n+1} = x_n - \frac{P(x_n)}{P'(x_n)}, \quad \text{where} \quad P'(x_n) = \frac{dP(x)}{dx} \quad \text{evaluated at} \quad x = x_n \quad \textbf{(8.4)}$$

NUMERICAL INTEGRATION

Numerical integration routines are extremely useful in almost all simulation programs, design of digital filters, theory of z-transforms, and almost any problem solution involving differential equations. And since digital computers have essentially replaced analog computers (which were almost true integration devices), the techniques of approximating integration are well developed. Several of the techniques are briefly reviewed here.

Euler s Method

For a simple first-order differential equation, say $dx/dt + ax = af$, one could write the solution as a continuous integral or as an interval-type one:

$$x(t) = \int^i [-ax(\tau) + af(\tau)]d\tau \quad \textbf{(8.5a)}$$

$$x(kT) = \int^{KT-T} [-ax + af]d\tau + \int_{kT-T}^{kT} [-ax + af]d\tau = x(kT-T) + A_{\text{rect}} \quad \textbf{(8.5b)}$$

Here, A_{rect} is the area of $(-ax + af)$ over the interval $(kT - T) < \tau < kT$. One now has a choice of looking back over the rectangular area or looking forward. The rectangular width is, of course, T. For the forward-looking case presented, a first approximation for x_1 is[2]

$$x_1(kT) = x_1(kT - T) + T[ax_1(kT - T) + af(kT - T)T$$
$$= (1 - aT)x_1(kT - T) + aTf(kT - T) \quad \textbf{(8.5c)}$$

Or, in general, for Euler's forward rectangle method, the integral may be approximated in its simplest form (using the notation $t_{\kappa+1} - t_\kappa$ for the width, instead of T which is $kT - T$) as

$$\int_{t_\kappa}^{t_{\kappa+1}} x(\tau) \, d\tau \approx (t_{\kappa+1} - t_\kappa)x(t_\kappa) \quad \textbf{(8.6)}$$

Trapezoidal Rule

The trapezoidal rule is based upon a straight line approximation between a function, $f(t)$, at t_0 and t_1. To find the area under the function, say, a curve, is to evaluate the integral of the function between point a and point b. The interval between these points is subdivided into subintervals; the area of each subinterval is approximated by a trapezoid between the end points. It will only be necessary to sum these individual trapezoids to get the whole area; by making the intervals all the same size, the solution will be simpler. For each interval of Δt (i.e., $t_{\kappa+1} - t_\kappa$), the area is then given by

$$\int_{t_\kappa}^{t_{\kappa+1}} x(\tau)\ d\tau \approx \frac{1}{2}(t_{\kappa+1} - t_\kappa)[tx(t_{\kappa+1}) + x(t_\kappa)] \tag{8.7}$$

This equation gives good results if the Δt's are small but it is for only one interval and the deviation from the actual value is called the "local error." This error may be shown to be $-\frac{1}{12}(\Delta t)^3 f''(t = \xi_1)$, where ξ_1 is between t_0 and t_1. For a larger "global error" it may be shown that

$$\text{Global error} = -\frac{1}{12}(\Delta t)^3\ [f''(\xi_1) + f''(\xi_2) + \cdots + f''(\xi_n)] \tag{8.8}$$

Following through on Eq. (8.8) allows one to predict the error for the trapezoidal integration. This technique is beyond the scope of this review or probably the examination; however, for those interested, refer to pages 249–250 of reference 1.

NUMERICAL SOLUTIONS OF DIFFERENTIAL EQUATIONS

The numerical solution technique presented here is based upon a first-order ordinary differential equation. However, the method may be extended to higher-ordered equations by converting them to a matrix of first-ordered ones.

Integration routines produce values of system variables at specific points in time and update this information at each interval of Δt as T ($\Delta t = T = t_{k+1} - t$). Instead of a continuous function of time, $x(t)$, the variable x will be represented by discrete values, $x_0, x_1, x_2, \ldots, x_n$. Consider a simple differential equation as before, as (based upon Euler's method)

$$dx/dt + ax = f(t)$$

Now assume the delta time periods, T, are fixed (not all routines use fixed step sizes), then one writes the continuous equation as a difference equation where $dx/dt \approx (x_{k+1} - x_k)/T = -ax_k + f_k$ or, solving for the updated value, x_{k+1},

$$x_{k+1} = x_k - Tax_k + Tf_k \tag{8.9a}$$

for fixed increments.

By knowing the first value of $x_{k=0}$ (or the initial condition), the solution may be achieved for as many "next values" of x_{k+1} as desired for some value of T. The difference equation may be programmed in almost any high-level language on a digital computer; however, T must be small compared to the shortest time constant of the equation (here, $1/a$).

The following equation [with the "f" term meaning a "function of" rather than as a "forcing function" term as it is used in Eq. (8.5a)] is a more general form of Eq. (8.9a). This equation is obtained by letting the notation (x_{k+1}) become $y[(k+1)\ \Delta t]$ and is written (perhaps somewhat more confusing) as

$$y[(k+1)\Delta t]y(k\Delta t) + \Delta t\ f[(y(k\Delta t), k\Delta t)] \tag{8.9b}$$

Reduction of Differential Equation Order

To reduce the order of a linear time-dependent differential equation, the following technique is used. For example, assume a second-order differential equation:

$$x'' + ax' + bx = f(t)$$

define $x = x_1$ and $x' = x_1' = x_2$. Then, $x_2' + ax_2 + bx_1 = f(t)$,

$$x_1' = x_2 \leftarrow \text{ by definition}$$
$$x_2' = -bx_1 + ax_2 + f(t)$$

This equation can be extended to higher-order systems and, of course, be put in a matrix form (called the state variable form). It can also easily be set up as a matrix of first-order difference equations for solving digitally.

Packaged Programs

Most currently available packaged simulation programs use algorithms not necessarily based upon Euler's methods but more advanced methods such as the Runge-Kutta method. Automatic variable-step-size methods like Milne's may also be used. However, as mentioned before, these routines are all built into the packaged programs and may be transparent to the user. The user of a specialized program may be without knowledge of the high-level language being employed (except for certain modifications).

GLOSSARY OF COMPUTER TERMS

Accumulators	Registers that hold data, addresses, or instructions for further manipulations in the ALU
Address bus	Two-way parallel path connecting processors and memory and containing addresses
AI	Artificial intelligence
Algorithm	A sequence of steps applied to a given data set that solves the given problem
Alphanumeric data	Data containing the characters a,b,c…z,0,1,2,…, 9
ALU	Arithmetic and logic unit
ASCII	American Standard Code for Information Interchange, 7 bits/character (pronounced AS-key)
Asynchronous	Form of communication in which message data transfer is not synchronous with the basic transfer rate, requiring start/stop protocol
Baud rate	Bits per second
BIOS	Basic input/output system
Bit	0 or 1
Buffer	Temporary storage device
Byte	8 bits
Cache memory	Fast look-ahead memory connecting processors with memory, offering faster access to often-used data
Channel	Logic path for signals or data
CISC	Complex instruction set control
Clock rate	Cycles per second
Control bus	Separate physical path for control and status information
Control unit	Processor that fetches and decodes instructions to control the operations of registers and ALU's
CPU	Central processing unit, primary processor
Data buffer	Temporary storage of data
Data bus	Separate physical path dedicated to data
Digital	Discrete level or valued quantification (vs. analog or continuous valued)
Duplex communication	Communication mode in which data are transmitted in both directions, but only in one direction at any one time

Dynamic memory	Storage that must be continually hardware-refreshed to remain
EBCDIC	Extended Binary Coded Decimal Interchange Code, 8 bits/character (pronounced EB-see-dick)
EPROM	Erasable programmable read-only memory
Expert systems	Programs with AI that learn rules from external stimuli
Floppy disk	Removable disk media in various sizes, $5\frac{1}{4}$", $3\frac{1}{2}$"
Flowchart	Graphical depiction of logic using shapes and lines
G-byte	Gigabyte, 1,073,741,824 or 2^{30} bytes
Half-duplex communication	Two-way communications path in which only one direction operates at a time (transmit or receive)
Handshaking	Communications protocol to start/stop data transfer
Hard disk	Disk that has nonremovable media
Hardware	Physical elements of a system
Hexadecimal	Numbering system (base 16) that uses 0–9, A, B,..., F
Hierarchical database	Database organization containing hierarchy of indexes/keys to records
I/O	Input/output devices such as terminal, keyboard, mouse, and printer
IR	Instruction register
K-bytes	Kilobytes, 1024 or 2^{10} bytes
LAN	Local area network
LIFO	Last In-First Out
LSI	Large-scale integration
Main memory	That memory seen by the CPU
M-bytes	Megabytes, 1,048,576 or 2^{20} bytes
Memory	Generic term for random access storage
Microprocessor	Computer architecture with CPU in one LSI chip
MODEM	Modulator-demodulator
MOS	Metal oxide semiconductor
Multiplexer	Device that switches several input sources one at a time to an output
Narrow band	Frequency domain reference to channel bandwidth versus baseband
Nibbles	4 bits
Nonvolatile memory	As opposed to volatile memory, does not need power to retain present state
Number systems	Method of representing computer data as human readable information
OCR	Optical character recognition
OS	Operating system
OS memory	Memory dedicated to the OS, not useable for other functions
Parallel interface	A character (8-bit) or word (16-bit) interface with as many wires as bits in interface plus data clock wire.
Parity	Method for detecting errors in data, 1 extra bit carried with data, even or odd to match the sum of one bit in a data stream
PC	Program counter, or personal computer
Peripheral devices	Input/output devices not contained in main processing hardware
Program	A sequence of computer instructions
PROM	Programmable read-only memory
Protocols	Established set of handshaking rules enabling communication
Pseudocode	An English-like way of representing structured programming control structures
RAM	Random access memory
Real time/Batch	Method of program execution: real-time implies immediate execution; batch mode is postponed until run with a group of related activities
Relational database	Database organization that relates individual elements to each other without fixed hierarchical relationships
RISC	Reduced instruction set computer
ROM	Read-only memory
Scratchpad memory	High-speed memory either in hardware or software
Sequential storage	Memory (usually tape) accessed only in sequential order (n, $n+1$,...)

Serial interface	Single data stream that encodes data by individual bit per clock period
Simplex communication	One-way communication
Software	Programmable logic
Stacks	Hardware memory organization implementing Last In-Last Out access
Static memory	Memory that does not require intermediate refresh cycles to retain state
Structured programming	Use of structured programming constructs such as Do-While, If-Then, Else
Synchronous	Communication mode in which data and clock are at same rate
Transmission speed	Rate at which data are moved in baud (bits per second, bps)
Virtual memory	Addressable memory outside physical address bus limits through use of memory mapped pages
Volatile memory	Memory whose contents are lost when power is removed
VRAM	Video memory
Wide-band	300-3300 Hz
Words	8, 16, or 32 bits
WYSIWYG	What you see is what you get
16-bit	Basic organization of data with 2 bytes per word
32-bit	Basic organization of data with 4 bytes per word
64-bit	Basic organization of data with 8 bytes per word
80386	Intel's microprocessor architecture based on 16-bit data address bus, extended virtual memory, and external math coprocessor
80486	Upgrade to 80386 incorporating math coprocessor within VLSI
80586	Intel's microprocessor architecture based on 16-bit data, 32-bit address bus

REFERENCES

1. Gerald, C. E., and Wheatley, P. O. *Applied Numerical Analysis*, 3rd ed. Addison-Wesley, 1985.
2. Franklin, G. F., and Powell, J. O. *Digital Control of Dynamic Systems*, Addison-Wesley, 1980, p. 55.

PROBLEMS

8.1   In spreadsheets, what is the easier way to write
B1 + B2 + B3 + B4 + B5?
a.   SUM(B1:B5)
b.   (B1..B5)SUM
c.   @B1..B5SUM
d.   @SUMB2..B5

8.2   The address of the cell located at column 23 and row C is
a.   23C
b.   C23
c.   C.23
d.   23.C

8.3   Which of the following is not correct?
a.   A CD-ROM may not be written to by a PC.
b.   Data stored in a batch processing mode is always up-to-date.
c.   The time needed to access data on a disk drive is the sum of the seek time, the head switch time, the rotational delay, and the data transfer time.
d.   The methods for storing files of data in secondary storage are sequential file organization, direct file organization, and indexed file organization.

8.4   Which of the following is false?
a.   Flowcharts use symbols to represent input, output, decision branches, process statements, and other operations.
b.   Pseudocode is an English-like description of a program.
c.   Pseudocode uses symbols to represent steps in a program.
d.   Structured programming breaks a program into logical steps or calls to subprograms.

8.5   In pseudocode using DOWHILE, the following is true:
a.   DOWHILE is normally used for decision branching.
b.   The DOWHILE test condition must be false to continue the loop.
c.   The DOWHILE test condition tests at the beginning of the loop.
d.   The DOWHILE test condition tests at the end of the loop.

8.6   A spreadsheet contains the following formulas in the cells:

	A	B	C
1		A1+1	B1+1
2	A1^2	B1^2	C1^2
3	SUM(A1:A2)	SUM(B1:B2)	SUM(C1:C2)

If 2 is placed in cell A1, what is the value in cell C3?
a.   12
b.   20
c.   8
d.   28

8.7 A spreadsheet contains the following:

	A	B	C	D
1		3	4	5
2	2	A$2		
3	4			
4	6			

If you copy the formula from B2 into D4, what is the equivalent formula in D4?
a. A$2
b. C4
c. C4
d. C$2

8.8 A processing system is processor-limited when performing sorting of small tables in memory. Which would speed up computations?
a. Adding more main memory
b. Adding virtual memory
c. Adding cache memory
d. Adding peripheral memory

8.9 A small PC processing system performing large (1 MB) matrix operations is currently I/O-limited since memory is limited to 1 MB. Which would speed up computations?
a. Adding more main memory
b. Adding virtual memory
c. Adding cache memory
d. Adding peripheral memory

8.10 Transmission protocol: Serial, asynchronous, 8-bit ASCII, 1 start bit, 1 stop bit, 1 parity bit, 9600 bps. How long will it take for a 1-kilobyte file to be transmitted through the link?
a. 0.85 seconds
b. 0.96 seconds
c. 1.07 seconds
d. 1.17 seconds

8.11 Transmission protocol: Serial, synchronous, 8-bit ASCII, 1 parity bit, 9600 bps. How long will it take for a 1-kilobyte file to be transmitted through the link?
a. 0.85 seconds
b. 0.96 seconds
c. 1.07 seconds
d. 1.17 seconds

8.12 The binary representation 10101 corresponds to which base-10 number?
a. 3
b. 16
c. 21
d. 10,101

8.13 For the base-10 number 30, the equivalent binary number is
a. 1110
b. 0111
c. 1111
d. 11110

8.14 On personal computers the data storage device with the largest storage capacity is most likely to be
a. 3½" diskette
b. hard disk
c. random access memory
d. 3½" HD diskette

8.15 The output value of M in Exhibit 8.15 is closest to
a. 0
b. 2
c. 8
d. 48

8.16 Pseudocode can best be described as
a. a simple letter-substitution method of encryption
b. an English-like language representation of computer programming
c. a relational operator in a database
d. the way data are stored on a diskette

8.17 The pseudocode that best represents Exhibit 8.17 is
a. INPUT X,Y
    WAGE = X × Y
    IF X less than 40 THEN
    WAGE = WAGE + OVERTIME PAY
    END IF
    OUTPUT WAGE
b. INPUT hours worked and hourly pay
    WAGE = hours worked × hourly pay
    IF hours worked greater than 40 THEN
    WAGE = WAGE + (hours worked – 40)(overtime wage supplement)
    END IF
    OUTPUT WAGE
c. INPUT hours worked and hourly pay
    WAGE = (hours worked – 40)(overtime rate) + 40(hourly pay)
    OUTPUT WAGE
d. PRINT hours worked and hourly pay
    WAGE = hours worked × hourly pay
    IF hours worked equals 40 THEN
    WAGE = hours worked × hourly pay + (hours worked – 40) × overtime pay
    END IF
    OUTPUT WAGE

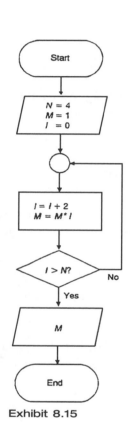

Start

$N = 4$
$M = 1$
$I = 0$

$I = I + 2$
$M = M * I$

$I > N?$ No

Yes

M

End

Exhibit 8.15

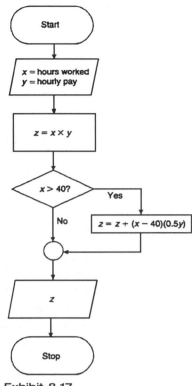

Exhibit 8.17

8.18 All but which of the following are true of spreadsheets?
   a. A cell may contain a label, value, formula, or function.
   b. A cell reference is made by a numbered column and lettered row reference.
   c. The cell content that is displayed is the result of a formula entered in that cell.
   d. Line graphs, bar graphs, stacked bar graphs, and pie charts are typical graphs created from spreadsheets.

SOLUTIONS

8.1 a. SUM(B1:B5) or @SUM(B1..B5).

8.2 b.

8.3 b. Batch mode processing always has delays in updating the database.

8.4 c. Pseudocode does not use symbols but uses English-like statements such as IF-THEN, and DOWHILE.

8.5 c. IF-THEN is normally used for branching. The DOWHILE test condition must be true to continue looping and the test is done at the beginning of the loop. The DOUNTIL test is done at the end of the loop.

8.6 b.   Plugging 2 into spreadsheet produces the following matrix:

	A	B	C
1	2	3	4
2	4	9	16
3	6	12	20

The value of C3 is 20.

8.7 d.   The formula contains a mixed reference. The "$" implies absolute row reference while the column reference is relative. The result of any copy would eliminate any answer except for an absolute reference to row 2. Relative column reference "A" gets replaced by "C." The cell contains C$2.

8.8 c.   Processor-limited sorting on small tables suggests either speeding up processor cycles or providing faster memory. Since speeding up the clock is not an option, then making memory faster is the answer. Adding more memory, or virtual memory does nothing for small tables. Only adding cache memory would allow the CPU to fetch recently used data without full memory cycles, thereby speeding up the sorting process.

8.9 a.   Processing is I/O-limited since all of matrix cannot fit into memory. Since this is a large matrix, cache memory probably would not effect processing. The best solution is adding more main memory to fit this problem entirely in memory.

8.10 d.

$$1 \text{ KB} = 2^{10} \text{ bytes} = 1024 \text{ bytes}$$
$$1024 \text{ bytes} + 3 \text{ overhead bits/byte}$$
$$= 1024(8 \text{ bits/byte} + 3 \text{ bits/byte overhead}) = 11{,}264 \text{ bits}$$

Minimum transmission time = 11,264 bits/(9600 bits/second) = 1.17 seconds.

8.11 b.

$$1 \text{ KB} = 2^{10} \text{ bytes} = 1024 \text{ bytes}$$
$$1024 \text{ bytes} + 1 \text{ overhead bit/byte}$$
$$= 1024(8 \text{ bits/byte} + 1 \text{ bit/byte overhead}) = 9216 \text{ bits}$$

Minimum transmission time = 9216 bits/9600 = 0.96 seconds.

8.12 c.   The binary (base-2) number system representation is

	2^4	2^3	2^2	2^1	2^0
	16	8	4	2	1
Binary number	1	0	1	0	1
	16	0	4	0	1 = 21

8.13 d.

2^5	2^4	2^3	2^2	2^1	2^0	
32	16	8	4	2	1	
		1	1	1	1	0 = 30

8.14 b.

8.15 d.

8.16 b.

8.17 b.

8.18 b.   A cell reference is made by a lettered column and a numbered row, example C3.

Legal and Professional Aspects: Ethics

Engineers, as members of a profession, are expected to conduct themselves in an ethical manner. This means that engineers must be aware of the standards of professional conduct (the ethical standards) and abide by them.

Codes of ethics have been prepared by the various national engineering societies, the engineering accreditation board (ABET), and the National Council of Examiners for Engineering and Surveying (NCEES). The *NCEES Model Rules of Professional Conduct* is reproduced in the Fundamentals of Engineering Supplied-Reference Handbook. Thus it is reasonable to assume that this code of ethics will be covered on the exam.

NCEES MODEL RULES OF PROFESSIONAL CONDUCT[1]

Section 240.15 Rules of Professional Conduct

I. Licensee's Obligation to Society
    a. Licensees, in the performance of their services for clients, employers, and customers, shall be cognizant that their first and foremost responsibility is to the public welfare.

---

[1] Reproduced by permission of NCEES, the copyright owner.

b. Licensees shall approve and seal only those design documents and surveys that conform to accepted engineering and surveying standards and safeguard the life, health, property, and welfare of the public.

c. Licensees shall notify their employer or client and such other authority as may be appropriate when their professional judgment is overruled under circumstances where the life, health, property, or welfare of the public is endangered.

d. Licensees shall be objective and truthful in professional reports, statements, or testimony. They shall include all relevant and pertinent information in such reports, statements, or testimony.

e. Licensees shall express a professional opinion publicly only when it is founded upon an adequate knowledge of the facts and a competent evaluation of the subject matter.

f. Licensees shall issue no statements, criticisms, or arguments on technical matters which are inspired or paid for by interested parties, unless they explicitly identify the interested parties on whose behalf they are speaking and reveal any interest they have in the matters.

g. Licensees shall not permit the use of their name or firm name by, nor associate in the business ventures with, any person or firm which is engaging in fraudulent or dishonest business or professional practices.

h. Licensees having knowledge of possible violations of any of these Rules of Professional Conduct shall provide the board with the information and assistance necessary to make the final determination of such violation. (*Section 150, Disciplinary Action, NCEES Model Law*)

II. Licensee's Obligation to Employer and Clients

a. Licensees shall undertake assignments only when qualified by education or experience in the specific technical fields of engineering or surveying involved.

b. Licensees shall not affix their signatures or seals to any plans or documents dealing with subject matter in which they lack competence, nor to any such plan or document not prepared under their direct control and personal supervision.

c. Licensees may accept assignments for coordination of an entire project, provided that each design segment is signed and sealed by the licensee responsible for preparation of that design segment.

d. Licensees shall not reveal facts, data, or information obtained in a professional capacity without the prior consent of the client or employer except as authorized or required by law.

e. Licensees shall not solicit or accept gratuities, directly or indirectly, from contractors, their agents, or other parties in connection with work for employers or clients.

f. Licensees shall make full prior disclosures to their employers or clients of potential conflicts of interest or other circumstances which could influence or appear to influence their judgment or the quality or their service.

g. Licensees shall not accept compensation, financial or otherwise, from more than one party for services pertaining to the same project, unless the circumstances are fully disclosed and agreed to by all interested parties.

h. Licensees shall not solicit or accept a professional contract from a governmental body on which a principal or officer of their organization serves as a member. Conversely, licensees serving as members, advisors, or employees of a government body or department, who are the principals or employees

of a private concern, shall not participate in decisions with respect to professional services offered or provided by said concern to the governmental body which they serve.
(*Section 150, Disciplinary Action, NCEES Model Law*)

III. Licensee's Obligation to Other Licensees

    a. Licensees shall not falsify or permit misrepresentation of their, or their associates', academic or professional qualifications. They shall not misrepresent or exaggerate their degree of responsibility in prior assignments nor the complexity of said assignments. Presentations incident to the solicitation of employment or business shall not misrepresent pertinent facts concerning employers, employees, associates, joint ventures, or past accomplishments.

    b. Licensees shall not offer, give, solicit, or receive, either directly or indirectly, any commission, or gift, or other valuable consideration in order to secure work, and shall not make any political contribution with the intent to influence the award of a contract by public authority.

    c. Licensees shall not attempt to injure, maliciously or falsely, directly or indirectly, the professional reputation, prospects, practice, or employment of other licensees, nor indiscriminately criticize other licensees' work.
(*Section 150, Disciplinary Action, NCEES Model Law*)

Example 9.1

The *NCEES Model Rules of Professional Conduct* allow an engineer to do which one of the following?

a. Accept money from contractors in connection with work for an employer or client.

b. Compete with other engineers in seeking to provide professional services.

c. Accept a professional contract from a governmental body even though a principal or officer or the engineer's firm serves as a member of the governmental body.

d. Sign or seal all design segments of the project as the coordinator of an entire project.

Solution

    b. Although the other items are not allowed by the *Model Rules*, nowhere does it say that an engineer cannot compete with other engineers in seeking to provide professional services. But, of course, he/she should compete in an ethical manner.

No code of ethics can cover every kind of situation an engineer will encounter. It does, however, describe in broad terms the ethical principles by which an engineer should be guided.

The examination asks ethics questions concerning relations with clients, peers, and the public. The *NCEES Model Rules of Professional Conduct* are organized into these same three categories. Thus the exam ethics questions likely are written to see if the examinee has read and generally understands the standards of professional conduct.

LEGAL AND CONTRACT TERMS

Terms

(a)   How many parties constitute the minimum number needed to form a contract?

(b)   What are the essential ingredients of a contract?

(c)   What is the difference between a *formal* and an *informal* contract?

(d)   What is meant by common law?

(e)   What is meant by statute law?

(f)   What is equity?

(g)   Define the terms *owner*, *engineer*, and *contractor* as they may appear in a contract document.

(h)   What is the distinction between contract plans and specifications?

(i)   When construction is performed by *force account*, what does this mean?

(j)   What is meant by the term *incorporate by reference*?

(k)   Why does one specify "or equal" or "an approved equivalent" in a specification for materials? Who does the approving?

(l)   Distinguish between the terms *liquidated damages* and *punitive damages*.

(m)   Describe a surety bond. What is its purpose?

De nitions

(a)   Contracts may be formed by two or more parties. That is, there must be a party to make an offer and a party to accept.

(b)   To be enforceable by law, a contract must contain the following essential elements:
   1. There must be a real agreement.
   2. The subject matter must be lawful.
   3. There must be a valid consideration.
   4. The parties must be legally competent.
   5. The contract must comply with the provisions of the law with regard to form.

(c)   A formal contract depends upon a particular form, or mode of expression, for legal efficacy. All other contracts are called informal contracts since they do not depend upon mere formality for their legal existence.

(d)   Common law comprises the body of those rules of action and principles which derive their authority solely from usage and customs.

(e)   Statute law consists of acts or statutes established by legislative action.

(f)   Equity is a system of doctrines supplementing common and statute law, for example, the Maxims of Equity.

(g)   The "owner" as specified is the principal. The "engineer" may be a party to an agency with the principal, or he may be an independent contractor. Control and responsibility must be clearly defined. The "contractor" may also be a party to an agency or an independent contractor.

(h) Specifications are written instructions that accompany and supplement plans. Whereas the drawings show the physical characteristics of the work, the specifications cover the quality of the materials, workmanship, and other technical requirements. The plans and specifications form and guide the standards of performance that will be required.

(i) When the owner elects to do work with his own forces, instead of employing a construction contractor, this is known as the force account method. Under this method, the owner maintains direct supervision over the work, furnishes all materials and equipment, and employs workmen on his own payroll.

(j) The term "incorporate by reference" refers to the practice of making a document legally binding by referencing it within a contract. The document becomes part of the legal contract even though it is not attached to or reproduced in the contract. This is necessary to eliminate unnecessary repetition.

(k) The terms "or equal" and "or an approved equivalent" are used in specifications for materials for several reasons. They take care of the problem of replacement if the original is not available. The terms also allow the bidder to bid lower if he has a supplier with different but equal material at lower cost. The engineer is the party who approves the replacement materials.

(l) Liquidated damages apply when a specific sum of money has been expressly stipulated as the amount of damages to be recovered by either party for a breach of the agreement by the other. Punitive damages are used to punish the defendant in certain situations involving willful, wanton, malicious, or negligent torts.

(m) Surety bonds are normally used in connection with competitive-bid contracts, namely, bid bonds, performance bonds, and payment bonds. The bonds are issued by a third party to guarantee the faithful performance of the contractor.

PROBLEMS

9.1 Jack, a consulting environmental engineer, was hired by the local garbage company to represent them in a hearing before the local town council. In preparing for an appearance before the town council, Jack worked at the garbage company truck storage yard. As he was starting to leave late one evening he discovered that a garbage company employee was moving garbage trucks to the edge of a stream embankment and dumping the liquid that came from the garbage into the stream. The next morning Jack spoke to the truck storage yard manager about what he had seen. The manager told Jack that they had always done it that way and for him to forget it.

Jack feels very uncomfortable with the situation; he knows the dumping is a clear violation of an environmental regulation. Consider Jack's four alternatives and select the best one.

a. Terminate his consulting with the garbage company.
b. Assume that since it has gone on for a long time there is no major hazard being created and forget what he saw.
c. At the next opportunity advise the local town council of the situation.
d. Write a letter to the garbage company president outlining what he discovered and advising the president that the dumping practice is an environmental regulation violation.

9.2 Dave's employer is a small engineering firm. For one of his first jobs as a young engineer Dave was assigned to write the specifications for a 10-inch pump and complete the purchase. He wrote the specifications and sent a sealed bid request to six firms. Only five firms submitted the bids before the bid request deadline. On the last day Dave called the sixth firm to remind them of the sealed bid request. The estimator indicated he was very busy and did not want to take the time to prepare a sealed bid unless he had a good chance of supplying the pump. Dave wants the sixth quote. Which course of acion should care take?

a. Since he doesn't plan to open the sealed bids rights away, Dave offers to let this last firm submit its quote a week late.
b. Dave opens the five sealed bids and reads them over the phone to the estimator, who says he thinks he could beat them and submit a bid by the sealed bid deadline later that day.
c. Dave already has five bids so he calls the sixth firm back and tells them there is nothing he can do for them.
d. Dave opens the five bids and finds the lowest one was $8750. He calls the estimator for the sixth firm and says, "I can't tell you the bids, but unless you can get under about $8800 forget about it."

9.3 Joan, a consulting soils engineer, submitted a draft of her report on soil conditions at a plant site to her client's chief engineer. The chief engineer read the draft of the report and indicated to Joan that he wanted several portions of the report changed so the planned building could be designed with a smaller, less expensive foundation. The chief engineer told Joan the allowable soil pressure value he wanted her to recommend in the final report.

Following this meeting Joan went back to her own office to decide what to do. She believes she has four alternatives. Which one should she select?
a. Submit the draft soil report as the final report without any changes.
b. Reexamine her field test data to see if the chief engineer's desired allowable soil pressure value could be recommended. If so, she would do it. If not, she would leave the report as it is.
c. Make the changes requested by the client's chief engineer and submit the final report.
d. Write to the president of this client's firm and describe the request of the chief engineer. Advise the president that the draft report is being submitted as the final report.

9.4   Jim, a newly hired engineer, was asked by his supervisor to see what he could learn about their competitors' future product plans. This is an important assignment and Jim is anxious to do a good job. He has devised four alternatives. Which one should he select?
a. Jim could call the competitors and tell them he is a college student working on a report for one of his classes. Jim thinks the competitors would be very forthright in this situation.
b. Jim could call up some college buddies who are now working for the competitors. A lunch together could easily result in trading some confidential product plans if Jim worked at it.
c. Jim could call the competitors and simply ask about their future product plans. If the competitors asked who Jim is, he will honestly tell them who his employer is. But if they do not ask, then Jim will not volunteer the information.
d. Jim will contact the competitors and at the beginning of the conversation explain who he is, his employer, and that he has been assigned to study competitors' future product plans.

9.5   When Glenn, a product development engineer, discovered that his manufacturing company was about to lose a major customer to an overseas competitor, he resolved to warn the customer that he might be making a serious mistake. He is considering what statement to make to the customer. Which adheres to ethical guidelines?
a. "I read recently that a lot of this kind of product is being assembled in the foreign country by prison labor. I hope you agree it is unethical to use prisoners in that way."
b. "We have some sales literature that describes features of our product and compares them with several competing products. I think you might find this information helpful, so I'm sending you a copy."
c. "I know a number of their engineers because we went to college together. Believe me, they probably have copied our product as I doubt they would be skillful enough to design one from scratch."
d. "My boss says the overseas competitor will probably give you a low bid to get the initial work, but you will pay a lot more on any subsequent orders."

SOLUTIONS

9.1 d. As a consultant to the garbage company Jack needs to see that responsible management is aware of his liquid dumping discovery. The appropriate next step is to advise the company president. Neither (a) nor (b) is satisfactory as neither are fails to take appropriate action when some action is clearly called for. Choice (c), on the other hand, is a premature action that may be a public disclosure prior to a thorough evaluation of all the facts.

9.2 c. In purchasing equipment using a sealed bid method, Dave has the responsibility to treat all bidders fairly. If the sixth bidder indicated he could submit the bid the morning following the stated bid deadline, this might be considered a reasonable request on this relatively small purchase. Thus Dave might consider it ethical and fair to allow this bidder a few additional hours to submit the quote. A substantial concession (like a week's delay) would not be appropriate unless it were made available to all bidders in a timely manner. Providing information about competing sealed bids to one bidder is unethical and unfair.

9.3 b. Alternative (c) is clearly inappropriate as it would not represent Joan's professional opinion. Both (a) and (d) might be suitable if Joan considers the chief engineer's desired allowable soil pressure value obviously wrong. The (b) alternative seems most practical. Soil analysis is not an exact science. A reevaluation of the field test data might indicate a somewhat larger allowable soil pressure value—possibly even the one the chief engineer wants. Thus the reevaluation and final report would be accurate and professional.

9.4 d. Jim immediately is faced with a situation where he must recognize what ethical conduct is. Alternatives (a) and (b) are unethical practices. Neither misrepresenting oneself to a competitor nor disclosing an employer's confidential data are appropriate. Between alternatives (c) and (d) the situation is less clear. An ethical engineer must proceed forthrightly, but must he take the path of alternative (d)?

Jim's problem may be his inexperience. Other alternatives are available, such as requesting all available sales literature of the competitors' products. By careful study, and by knowing the basic manufacturing and development capabilities, one can usually correctly predict the direction of products. Further, since this evaluation would naturally help decide the direction of your own product line, a more careful market analysis of your own situation is certainly warranted. Often manufacturers purchase competitive products and benchmark their manufacturing technology. Since this is an ethics test question, the safest answer is probably best.

9.5 b. Choices (a), (c), and (d) represent malicious comments attempting to injure a competitor's reputation. Answer (b) represents material that has been examined within Glenn's company and has been approved for distribution. Since it was intended for customers or prospective customers, Glenn would appear to be acting in an appropriate and ethical manner if he selects this alternative.

Afternoon Sample Examination

Topic	Number of Problems
COMPUTER & NUMERICAL METHODS	6
CONSTRUCTION MANAGEMENT	3
ENVIRONMENT ENGINEERING	6
HYDRAULICS & HYDROLOGIC SYSTEMS	6
LEGAL & PROFESSIONAL ASPECTS	3
SOIL MECHANICS & FOUNDATIONS	6
STRUCTURAL ANALYSIS	6
STRUCTURAL DESIGN	6
SURVEYING	6
TRANSPORTATION FACILITIES	6
WATER PURIFICATION & TREATMENT	6

INSTRUCTIONS FOR AFTERNOON SESSION

1. You have four hours to work on the afternoon session. You may use the *Fundamentals of Engineering Supplied-Reference Handbook* as your *only* reference. Do not write in this handbook.

2. Answer every question. There is no penalty for guessing.

3. Work rapidly and use your time effectively. If you do not known the correct answer, skip it and return to it later.

4. Some problems are presented in both metric and English units. Solve either problem.

5. Mark your answer sheet carefully. Fill in the answer space completely. No marks on the workbook will be evaluated. Multiple answers receive no credit. If you make a mistake, erase completely.

Work 60 afternoon problems in four hours.

FUNDAMENTALS OF ENGINEERING EXAM

AFTERNOON SESSION

(A) (B) (C) (D) Fill in the circle that matches your exam booklet

1. (A) (B) (C) (D)	16. (A) (B) (C) (D)	31. (A) (B) (C) (D)	46. (A) (B) (C) (D)
2. (A) (B) (C) (D)	17. (A) (B) (C) (D)	32. (A) (B) (C) (D)	47. (A) (B) (C) (D)
3. (A) (B) (C) (D)	18. (A) (B) (C) (D)	33. (A) (B) (C) (D)	48. (A) (B) (C) (D)
4. (A) (B) (C) (D)	19. (A) (B) (C) (D)	34. (A) (B) (C) (D)	49. (A) (B) (C) (D)
5. (A) (B) (C) (D)	20. (A) (B) (C) (D)	35. (A) (B) (C) (D)	50. (A) (B) (C) (D)
6. (A) (B) (C) (D)	21. (A) (B) (C) (D)	36. (A) (B) (C) (D)	51. (A) (B) (C) (D)
7. (A) (B) (C) (D)	22. (A) (B) (C) (D)	37. (A) (B) (C) (D)	52. (A) (B) (C) (D)
8. (A) (B) (C) (D)	23. (A) (B) (C) (D)	38. (A) (B) (C) (D)	53. (A) (B) (C) (D)
9. (A) (B) (C) (D)	24. (A) (B) (C) (D)	39. (A) (B) (C) (D)	54. (A) (B) (C) (D)
10. (A) (B) (C) (D)	25. (A) (B) (C) (D)	40. (A) (B) (C) (D)	55. (A) (B) (C) (D)
11. (A) (B) (C) (D)	26. (A) (B) (C) (D)	41. (A) (B) (C) (D)	56. (A) (B) (C) (D)
12. (A) (B) (C) (D)	27. (A) (B) (C) (D)	42. (A) (B) (C) (D)	57. (A) (B) (C) (D)
13. (A) (B) (C) (D)	28. (A) (B) (C) (D)	43. (A) (B) (C) (D)	58. (A) (B) (C) (D)
14. (A) (B) (C) (D)	29. (A) (B) (C) (D)	44. (A) (B) (C) (D)	59. (A) (B) (C) (D)
15. (A) (B) (C) (D)	30. (A) (B) (C) (D)	45. (A) (B) (C) (D)	60. (A) (B) (C) (D)

DO NOT WRITE IN BLANK AREAS

SAMPLE EXAM

1. A soil specimen has a void ratio of 0.6, moisture content of 10%, and $G = 2.67$. Determine the moist unit weight.
   a. 13.4 kN/m^3
   b. 15.6 kN/m^3
   c. 18 kN/m^3
   d. 20.1 kN/m^3

2. Calculate the seepage loss through the 0.5-m thick sandy silt layer below the earth dam shown (m^3/hr/m). Given: coefficient of permeability, $k = 3.66$ cm/hr.
   a. 0.0023
   b. 0.0037
   c. 0.0051
   d. 0.0093

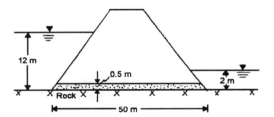

3. A normally consolidated clay layer having a thickness of 3 m is shown. A laboratory consolidation test on the same clay gave the following results:

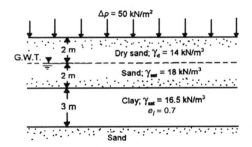

p (kN/m^2)	e
75	0.0661
150	0.557

   The settlement of the clay layer due to a surcharge $\Delta p = 50$ kN/m^2 is closest to:
   a. 0.14 m
   b. 0.25 m
   c. 0.56 m
   d. 0.75 m

4. A silty clay layer of 6-m thickness is shown on the following page. It has a drained friction angle of 21° and a cohesion of 20 kN/m^2. What would be the drained shear strength at A (in kN/m^2)?
   a. 30
   b. 45
   c. 60
   d. 75

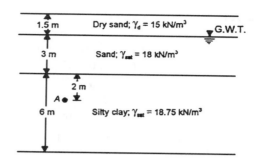

5. For a granular soil, the dry unit weight in the field is 16.35 kN/m³. The maximum and minimum dry unit weights of the same soil as determined in the laboratory are 17.6 kN/m³ and 14.46 kN/m³, respectively. Given $G = 2.66$, the relative density of compaction in the field is closest to

  a. 0.48
  b. 0.53
  c. 0.65
  d. 0.76

6. For a clayey soil, given:
Plasticity index =18
Plastic limit = 15
Specific gravity of soil solids, G = 2.68
Determine the void ratio of the soil at liquid limit. Note: At liquid limit, the soil is saturated.

  a. 0.60
  b. 0.70
  c. 0.80
  d. 0.90

7. The ratio of width b to depth y that maximizes the discharge through a fixed cross-sectional area in an open channel of rectangular section is

  a. $b/y = \dfrac{1}{3}$
  b. $b/y = \dfrac{1}{2}$
  c. $b/y = 1$
  d. $b/y = 2$

8. The depth of flow in a concrete-lined ($n = 0.012$) triangular channel, laid on a slope of 2 m per 3 km and carrying a discharge of 1.0 m³/s, is closest to

  a. 0.85 m
  b. 0.91 m
  c. 0.97 m
  d. 1.03 m

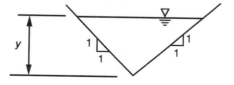

9. The hydraulic radius for a circular pipe of diameter D flowing half-full is
a. $D/4$
b. D
c. $\pi D/2$
d. πD

10. The troweled concrete channel shown conveys water at a depth of 1 m on a slope of 0.0004. The discharge is closest to
a. $1.0 \text{ m}^3/\text{s}$
b. $2.1 \text{ m}^3/\text{s}$
c. $3.2 \text{ m}^3/\text{s}$
d. $10.0 \text{ m}^3/\text{s}$

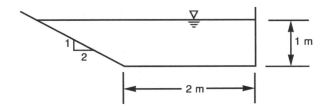

11. A 2-m-diameter pipe connecting two reservoirs has an interior surface of smooth masonry ($C = 120$) and is 5000 m long. The difference in reservoir water surface elevations is 6.0 m. According to the Hazen-Williams formula, the discharge is nearest to
a. $1.7 \text{ m}^3/\text{s}$
b. $2.7 \text{ m}^3/\text{s}$
c. $5.5 \text{ m}^3/\text{s}$
d. $8.5 \text{ m}^3/\text{s}$

12. A new cast iron pipe ($C = 130$) is to convey $4.5 \text{ m}^3/\text{s}$ a distance of 6 km from one reservoir to another reservoir whose water surface is 7.5 m lower than the first reservoir. The most appropriate pipe diameter is
a. 1.2 m
b. 1.8 m
c. 2.4 m
d. 3.0 m

QUESTIONS 13-14

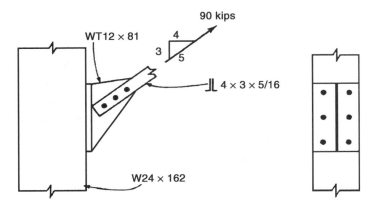

The relevant properties of one angle are $A_g = 2.09$ inches^2, hole diameter $= d_b + 0.125$ inches. The allowable stress may be increased by one-third when considering wind loads.

13.   All bolts in the wind bracing connection shown in the figure are $\frac{7}{8}$-inch A325 slip-critical bolts in standard holes ($A_g = 2.09$). All structural sections are Grade A36 steel. The effective net area of the angle brace is most nearly:
   a.   2.0 square inches   c.   3.0 square inches
   b.   2.5 square inches   d.   3.5 square inches

14.   The capacity of the angle brace based on its effective net area is most nearly:
   $F_u = 58$, $A_e = 3.02$
   a.   101 kips   c.   111 kips
   b.   106 kips   d.   116 kips

15.   The W14 × 48 Grade A36 column shown below supports an axial load only. The relevant properties of the W14 × 48 are $A_s = 14.1$ inches^2, $E_s = 29,000$ ksi, $F_y = 36$ ksi, $r_x = 5.85$ inches, and $r_y = 1.91$ inches. The allowable axial load is most nearly:
   a.   230 kips
   b.   240 kips
   c.   250 kips
   d.   260 kips

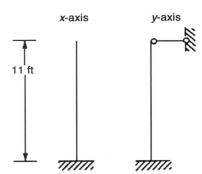

16.   A 500-mm diameter, spirally reinforced column with 40 mm cover to the spiral is reinforced with 10 No. M30 bars. The concrete strength is 35 MPa and the tensile strength of the reinforcement is 400 MPa. The column may be classified as a short column.
   The column supports only axial load and the design axial load strength is given most nearly by:
   a.   4800 kN   c.   5200 kN
   b.   5000 kN   d.   5400 kN

17.   For the column in Problem 16, the minimum permitted diameter of the spiral reinforcement is:
   a. $\frac{1}{4}$-inch   c. $\frac{1}{2}$-inch
   b. $\frac{3}{8}$-inch   d. $\frac{5}{8}$-inch

18. For the column in Problem 16, is the reinforcement ratio of the column satisfactory?
    a. yes
    b. no

19. A 300-m-radius simple curve with a tangent deflection of 40° has been staked on the ground. It is desired to **sharpen** the curve so as to move the midpoint of the curve a distance of 3 m **radially**. The tangent deflection must remain fixed at 40° so the direction of the tangents is not changed. The radius (m) of the new curve is most nearly:
    a. 297
    b. 331
    c. 253
    d. 340

20. An existing vertical curve joins a +4% with a –4% grade. Assume $f = 0.4$, a perception-reaction time of 2.5 s, and that $S < L$. If the length of the curve is 75 m, the maximum safe speed (km/h) on the curve is most nearly:
    a. 50
    b. 40
    c. 60
    d. 45

QUESTIONS 21-22

Given the following traffic count data:

Time Interval	No. of Vehicles
5:00–5:15	1000
5:15–5:30	1100
5:30–5:45	1200
5:45–6:00	900

21. The peak hour factor is most nearly:
    a. 0.950
    b. 0.875
    c. 0.050
    d. 0.750

22. The peak rate of flow (veh/hr) is most nearly:
    a. 3600
    b. 4400
    c. 4800
    d. 3675

QUESTIONS 23-24

Given the following spiral and simple curve data: PI at station 1 + 086.271, $R = 300$ m, $\Delta = 16$ degrees, $\lambda_s = 52.083$ m, $p = 0.377$ m, $k = 26.035$ m.

23.   Assuming a rate of increase of centripetal force (C) of 0.6 m/s^3, the recommended design speed (km/h) for the curve is most nearly:
   a.   100
   b.   60
   c.   120
   d.   75

24.   The stationing of the TS for the curve is most nearly:
   a.   1 + 018
   b.   1 + 154
   c.   1 + 138
   d.   1 + 000

25.   A section of highway has the following design requirements.
   Design speed = 80 km/hr
   Coefficient of friction = 0.35
   Perception-reaction time = 2.5 s
   Grade = +3%
   Driver eye height = 1070 mm
   Object height = 150 mm

   The stopping sight distance (m) for this highway is most nearly:
   a.   135
   b.   120
   c.   130
   d.   110

26.   The areas (in m^2) of the cut sections of a proposed highway are shown below.

Stations	Area, m^2
5 + 000	27.9
5 + 050	326.1
5 + 100	965.3
5 + 200	1651.8
5 + 300	2126.6

   The total volume of cut (to the nearest cubic meter) between stations 5 + 000 and 5 + 100 is most nearly (use the average end area method):
   a.   35,615
   b.   34,865
   c.   44,420
   d.   41,135

27.   A 11-m-wide vertical wall roadway tunnel is to be constructed using the cut and cover method. The cross sections to be excavated are shown below.
       The volume of earth in cubic meters to be excavated between stations 1 + 130 and 1 + 260 is most nearly (use the prismoidal method):
   a.   5165
   b.   5585
   c.   7860
   d.   2340

Station	Area, m^2
1 + 000	20
1 + 100	30
1 + 130	35
1 + 200	40
1 + 260	45
1 + 300	40
1 + 400	35

28. The binary representation 10101 corresponds to which base-10 number?
  a. 3
  b. 16
  c. 21
  d. 10,101

29. For the base-10 number 30, the equivalent binary number is
  a. 1110
  b. 0111
  c. 1111
  d. 11110

30. On personal computers the data storage device with the largest storage capacity is most likely to be
  a. $3\frac{1}{2}$" diskette
  b. hard disk
  c. random access memory
  d. $3\frac{1}{2}$" HD diskette

31. Pseudocode can best be described as
  a. a simple letter-substitution method of encryption
  b. an English-like language representation of computer programming
  c. a relational operator in a database
  d. the way data are stored on a diskette

32. All but which of the following is true of spreadsheets?
  a. A cell may contain a label, value, formula, or function.
  b. A cell reference is made by a numbered column and lettered row reference.
  c. The cell content that is displayed is the result of a formula entered in that cell.
  d. Line graphs, bar graphs, stacked bar graphs, and pie charts are typical graphs created from spreadsheets.

33. The following polynomial obviously has three roots:

$$P(x) = x^3 + 4x^2 + 6x + 4$$

Is $x = -1$ one of the exact roots?
  a. No, because its less than the last coefficient.
  b. Yes, because there is a remainder when using synthetic division.
  c. No, because there is a remainder when using synthetic division.
  d. Yes, because there is no remainder when using synthetic division.

34. For the polynomial in Problem 33, if a second trial root is to be found (using Newton's method for finding roots), what is this next trial division?
a. 1
b. 2
c. −1
d. −2

35. For the differential equation $dx/dt + 5x = 1$, $x(0) = 0$; what is the value of x suggested by Euler's method after the first iteration after the initial condition?
a. 0.1
b. 0.2
c. 0.3
d. 0.4

36. The *NCEES Model Rules of Professional Conduct* do not allow an engineer to improve his/her competence in a specific area in which one of the following ways?
a. Studying the work of other engineers.
b. Studying at a university outside of the United States.
c. Performing services in an area where the engineer is not yet competent.
d. Working in a specific area under the direction of a competent but unregistered engineer.

37. The *NCEES Model Rules of Professional Conduct* require the registered engineer to do all but one of the following actions. Which one should not be done?
a. Make no statements on technical matters unless any interested party for whom the engineer is acting is disclosed.
b. Approve documents only if they safeguard the life, health, property, and welfare of the public.
c. Reveal facts and data obtained in a professional capacity only after obtaining consent of the client or employer, unless required to reveal the facts and data by law.
d. Comment on professional matters publicly even though the engineer has not yet had an opportunity to make a careful evaluation of the facts and subject matter.

38. The *NCEES Model Rules of Professional Conduct* prohibit the registered engineer from doing all but one of the following. Which one can the engineer do?
a. Indiscriminately criticize the work of another registered engineer.
b. Give gifts to people in order to promote a business and obtain work.
c. Affix a signature or seal to plans or documents not prepared under the engineer's control and supervision.
d. Accept compensation from more than one party for services on the same project, even though the circumstances are fully disclosed and agreed to by all parties.

39. Jane Doe is a self-employed consulting engineer in the engineering department of ABC company where you work as a full-time staff process engineer. One of your expertises is relief system design. Things are slow, and even though you are well recognized, you are facing a potential layoff. She approaches you for part-time help, which may be summarized by the following conversation:

Jane: "Hey Phil, how about earning some extra money, say forty dollars per hour, from XYZ Company, for helping in my calculations to check the size of relief devices on weekends? I will do all the leg work, collect all the pertinent data, all you have to do is to check whether the device is OK or not OK, and show the basis of your calculations."

Phil: "Is it not illegal to have a part-time job when I am a full-time employee?"

Jane: "I don't think so. I know John, your project manager, has a restaurant business. Roger, your own colleague, helps in income tax preparation on weekends."
What would you do?

a. Check the employment agreement to make sure nothing was signed against accepting a part-time job, accept the offer, work at home without using any company time or resources, and follow "don't-ask-don't-tell" principle.

b. Accept the offer, work at the office outside working hours, and use the company's method and copier.

c. Read the company ethics book and guidelines to check whether there is any clear definition of conflict of interest. If the definition is clear, reject the offer. If unclear, ask the Human Resources Department for permission by a short memo, and then decide.

d. Invite Jane and your boss for a dinner, casually refer to the situation at an appropriate moment, have his permission well heard so that Jane remains a witness, and then accept the offer.

40. Suppose you work for a large electrical design firm that designs electrical control systems for industrial manufacturing sites. Assume your design requirements include specifying "short circuit and ground fault protection" for motor circuits. These device specifications should comply with the National Electric Code. Your design, started in early 1991, involved the selection of a non–time delay fuse for a three-phase, 230-V, 50-hp, 115-A (its nameplate values) induction motor for a molded rubber parts plant. The code book states that the full rated current for this motor is to be 130 amperes. The book also states that for this type of non–time delay fuse the current should not be more than 300% of rated. However, you find that commercially available fuses come only in increments of 50 amperes near that value. You choose the 400-amp fuse since the code implies that it is permissible (and widely accepted practice) to "round up" to this higher value. Your design was not implemented as the project was tentatively canceled due to lack of funding. In 1995 the exact same project was funded and the project went ahead. Assume you have gone on to work on other projects. You are aware that in the 1993 edition of this handbook a number of revisions were made, one requiring fusing and circuit breakers to use the next lower, or "rounded down", fuse and circuit breaker sizes. What action, if any, should you take?

a. Ignore the project since you are working in another area now.
b. Call the electrical engineer you know is working on the project to remind her that the prior work was based on the code in force in 1991 and not the latest code.
c. Write a memo to the project supervisor about the change in the code and the need to revise the fuse rating.
d. The client for the project, the firm that produces molded rubber parts, has a project manager overseeing the plant design. Call this project manager and tell him of the needed change in fuse rating to match the present electrical code.

41. A certain technology company, located in California, produces a very popular electronic device that emits an amount of radiation that is within established limits. The company sells the product both domestically and abroad. The company is able to increase its production by buying some of the required components overseas. You are the supervisor of the company's test laboratory, which samples and tests for radiation levels. It is a standard and acceptable practice to sample a certain percentage of the devices produced to check for compliance. After a large production run, your sampling procedure shows that the limits of radiation have been exceeded. You advise your company's managers of this higher radiation level, which is due to variation in the overseas components (that is, their quality control is poor). A meeting is called between yourself, two other engineers, and two plant managers. The cause of the problem is the variation in the overseas components. The possible courses of action discussed are

a. The tolerances of a few of the sampled components are usually detectable. To create a distorted testing record, selectively choose the samples to test. Because the test records will show the percent sampled components to be within limits, ignore the problem (there are not that many devices too far out of specs).
b. Curtail the sales in the United States and sell to the overseas countries that don't have the radiation requirements.
c. Stop all sales until a new supplier is founded for higher-quality components, and attempt to modify existing devices to meet sample specifications. This action may delay an impending cost of living salary increase for most of the plant's employees.
d. Remove the label on each of the devices that states that the device meets all radiation requirements, reduce the price to remove the inventory, then phase in newer devices made with higher-quality components.

Each member of the committee is highly respected for his judgment, and the management wants honest opinions. Four of the members have voted, one for each of the possible courses of action, your vote then will be the deciding one. Which will you vote for?

QUESTIONS 42-43

A water has the following ionic composition:

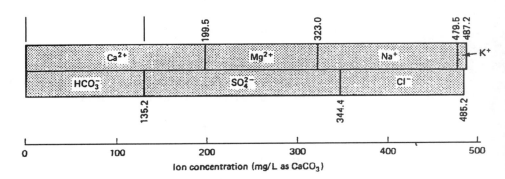

Ion concentration (mg/L as CaCO₃)

42. The total carbonate hardness as $CaCO_3$ is
a. less than 100 mg/L
b. greater than the noncarbonate hardness
c. only associated with Ca^{2+}
d. 244.4 mg/L

43. The $Ca(OH)_2$ needed to remove the carbonate hardness (CH) is
a. 1mol/mol CH
b. 2.2 mg/mg CH
c. 0.46 mg/mg CH
d. dependent upon both the calcium and magnesium concentrations

44. A single source of BOD to a river causes a minimum DO downstream of 6 mg/L. The initial DO deficit is zero and DO saturation is 10 mg/L
a. without aeration, the deficit will decrease
b. doubling the BOD load would double the deficit
c. doubling the BOD load would reduce the DO to 3 mg/L
d. doubling the BOD load would move the point where the deficit occurs closer to the discharge

45. A distilled water solution has 0.001 M H_2SO_4. The pH is
a. 2.7
b. 11.3
c. neutral
d. 3.0

46. Magnesium in an aqueous solution is in equilibrium with precipitated magnesium carbonate.
1. The magnesium concentration is dependent upon the pH.
2. The magnesium concentration is dependent upon the carbonate concentration.
3. The magnesium concentration is dependent upon the bicarbonate concentration.
4. The magnesium concentration is dependent upon alkalinity.

Which of the above statements are correct?
a. Statements 1 and 4 are correct.
b. Statements 2 and 3 are correct.
c. Statements 1 and 2 are correct.
d. All of the statements are correct.

47. In using chlorine for disinfection
a. concentration of HOCl is pH-dependent
b. [HOCl] equals [OCl⁻] at pH 7
c. effectiveness increases with time of contact
d. total elimination of bacteria is possible

48. A 61-cm-diameter circular concrete sewer is set on a slope of 0.001. The flow when it is half-full is
a. $\frac{1}{2}$ of flow at Q_{full}
b. 0.08 m³/s
c. 0.20 m³/s
d. 0.28 m³/s

49. The carbonaceous oxygen demand of 200 mg/L of glycine (H_2NCH_2COOH) is
a. less than the nitrogenous oxygen demand
b. depends on the reaction rate
c. greater than 120 mg/L
d. greater than 100 mg/L

50. A completely mixed lagoon has a waste flow of 1 m³/s of a BOD waste with a decay coefficient of 0.3/d and a concentration of 100 mg/L. If the effluent BOD must be 20 mg/L or less, then
a. the volume required is about 1×10^6 m³
b. the volume required is about 2×10^6 m³
c. the volume required is about 0.5×10^6 m³
d. an increase in temperature would increase the volume required

51. A completely mixed reactor with cell recycle is designed to treat a municipal waste. Assuming removal kinetics follow the equation:

$$r_{su} = -\frac{k \times S}{(K_s + S)}$$

which of the following statements is correct?
a. The effluent substrate concentration decreases with an increase in θ_c.
b. The food:microorganism level is independent of θ_c.
c. Microbe concentrations will be smaller than in the no-recycle case.
d. θ_c is independent of effluent quality.

52. Which of these statements are correct?
1. Chlorination of wastewater effluents requires more chlorine than chlorination of drinking water.
2. Chlorination of wastewater effluents requires three moles of chlorine for each mole of ammonia.

3. Chlorination of wastewater effluents is used to improve effluent quality.
4. Chlorination of wastewater effluents oxidizes other chemicals such as ferrous iron.

a. All of the above statements are correct.
b. None of the above statements is correct.
c. Only statements 1 and 3 are correct.
d. Only statements 1, 3, and 4 are correct.

53. A chemical is removed in a completely mixed reactor at a zero-order rate of 0.25 mg/L-d. The influent concentration is 10 mg/L and the hydraulic detention time is 10 days. For this reactor, which of these statements are correct?
1. The effluent concentration would be the same as that for a batch reactor with a 2-day retention time.
2. The effluent concentration will be about 7.5 mg/L.
3. The effluent quality would improve with longer detention times.
4. The effluent quality would improve with increased mixing.

a. All of the above statements are correct.
b. None of the above statements is correct.
c. Only statements 1 and 2 are correct.
d. Only statements 1, 3, and 4 are correct.

54. The load, W, at which the crane will tip is
a. more than 33.3 but less than 40 kips
b. more than 40 kips
c. more than 9.1 but less than 33.3 kips
d. none of the above

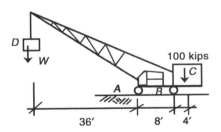

55. The force in bar EF is
a. 21.2 kips compressive
b. 21.2 kips tensile
c. 10.6 kips tensile
d. 10.6 kips compressive

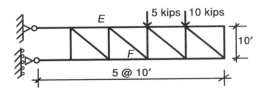

56.   The reaction at B is
   a.   23 kips up and to left
   b.   23 kips up and to right
   c.   23 kips down and to left
   d.   23 kips down and to right

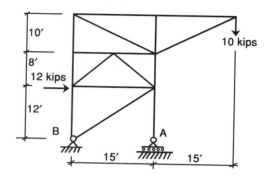

57.   If the force in bar BC is 28.57 kips tensile, the reaction at A must be
   a.   8.6 kips down
   b.   8.6 kips up
   c.   18.6 kips up
   d.   18.6 kips down

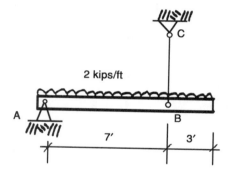

58.   The shearing force at section A–A is (ignore thickness of members)
   a.   12.8 kips
   b.   2.8 kips
   c.   5.0 kips
   d.   none of these

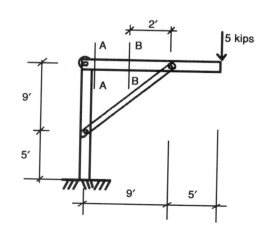

59. In the structure shown in Problem 58, the bending moment at section B–B is
   a. 50.5 kip-ft
   b. 19.4 kip-ft
   c. 12.0 kip-ft
   d. 15.4 kip-ft

60. A concrete-lined open channel with a slope of 0.34% has a bottom width of 1.5 m and side sloped of 2:1. This channel is to be designed to handle peak discharge of 15 m^3/s. Assume $n = 0.015$. The depth of flow in feet is nearest to:
   a. 1.0 m
   b. 1.2 m
   c. 1.4 m
   d. 1.6 m

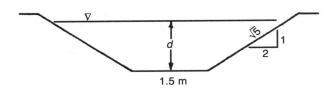

1.5 m

End of Exam. Please check your work.

SOLUTIONS

1.   c. Dry unit weight,

$$\gamma_d = \frac{G\gamma_w}{1+e} = \frac{(2.67)(9.81)}{1+0.6} = 16.37 \text{ kN/m}^3$$

Moist unit weight,

$$\gamma = \gamma_d(1+w) = (16.37)(1+0.1) = 18 \text{ kN/m}^3$$

2.   b. $Q = kiA$

$$k = 3.66 \text{ cm/hr}; \; i = \frac{12-2}{50} = 0.2; \; A = (0.5)(1) = 0.5 \text{ m}^2$$

$$Q = \left(\frac{3.66}{100}\right)(0.2)(0.5) = 0.0037 \text{ m}^3/\text{hr/m}$$

3. a.

$$C_c = \frac{e_1 - e_2}{\log\left(\dfrac{p_2}{p_1}\right)} = \frac{0.661 - 0.577}{\log\left(\dfrac{150}{70}\right)} = 0.279$$

Effective overburden pressure at the middle of the clay layer,

$$p_i = (14 \times 2) + (18 - 9.81)(2) + \frac{3}{2}(16.5 - 9.18) = 54.42 \text{ kN/m}^2$$

$$\Delta H = \frac{C_c H}{1 + e_i} \log\left(\frac{p_i + \Delta p}{p_i}\right) = \frac{(0.275)(3)}{1 + 0.7} \log\left(\frac{54.42 + 50}{54.42}\right)$$
$$= 0.139 \text{ m} = 139 \text{ mm}$$

4. b.

$s = c + \sigma' \tan\phi$
$c = 20 \text{ kN/m}^2; \ \phi = 21; \ \sigma' = (15)1.5 + 3(18 - 9.81) + 2(18.75 - 9.81)$
$\quad = 64.95 \text{ kN/m}^2$
$s = 20 + 64.95 \tan 21° = 44.9 \text{ kN/m}^2$

5. c.

$$D_d = \frac{\dfrac{1}{\gamma_{min}} - \dfrac{1}{\gamma_d}}{\dfrac{1}{\gamma_{min}} - \dfrac{1}{\gamma_{max}}} = \frac{\dfrac{1}{14.46} - \dfrac{1}{16.35}}{\dfrac{1}{14.46} - \dfrac{1}{17.6}} = 0.645 = 64.5\%$$

6. c. LL = PL + PI = 15 + 18 = 33
Degree of saturation,

$$S(\%) = \frac{wG}{e} \times 100$$

If $S = 100\%$,

$$e_{LL} = wG = \left(\frac{30}{100}\right)(2.68) = 0.804$$

7. d. The Manning discharge formula is

$$Q = \frac{1}{n} A R^{\frac{2}{3}} S^{\frac{1}{2}}$$

For a fixed area A the discharge is maximized when R is maximized. With $A = by$ and the wetted perimeter $p = b + 2y$,

$$R = \frac{A}{P} = \frac{by}{b + 2y} = \frac{A}{\dfrac{A}{y} + 2y}$$

$$\frac{dR}{dy} = \frac{-A}{\left(\dfrac{A}{y} + 2y\right)^2}\left(-\frac{A}{y^2} + 2\right) = 0$$

or $A = 2y^2 = by$ so $b = 2y$, $b/y = 2$.

8. c. The discharge is

$$Q = \frac{1}{n} A R^{\frac{2}{3}} S^{\frac{1}{2}}$$

where

$$A = y^2$$
$$P = 2\sqrt{2}\, y$$

$$R = A/P = \frac{y}{\left(2\sqrt{2}\right)}$$

Hence

$$1.0 = \frac{1}{0.012} y^2 \left(\frac{y}{2\sqrt{2}}\right)^{\frac{2}{3}} \left(\frac{2}{3000}\right)^{\frac{1}{2}}$$

or $y^{\frac{8}{3}} = 0.930$, which yields $y = (0.930)^{\frac{3}{8}} = 0.973$ m.

9. a. The hydraulic radius is $R = A/P$. Using $r = D/2$, for a half-full circular pipe one has $A = \pi r^2/2$, $P = \pi r$ so that

$$R = \frac{A}{P} = \frac{\frac{\pi r^2}{2}}{\pi r} = \frac{r}{2} = \frac{D}{4}$$

10. c. The discharge is

$$Q = \frac{1}{n} A R^{\frac{2}{3}} S^{\frac{1}{2}}$$

From the table of Manning n values, $n = 0.013$. Also,

$$A = 2 + \frac{1}{2}(1)(2) = 3.0 \text{ m}^2$$
$$P = 1 + 2 + \sqrt{5} = 5.24 \text{ m}$$

Therefore

$$Q = \frac{1}{0.013}(3.0)\left(\frac{3.0}{5.24}\right)^{\frac{2}{3}}(0.0004)^{\frac{1}{2}}$$

and

$$Q = 3.2 \text{ m}^3/\text{s}$$

11. c. For smooth masonry the Hazen-Williams coefficient is $C = 120$. The Hazen-Williams formula is

$$V = 0.849 C R^{0.63} S^{0.54}$$

Here

$$A = \frac{\pi}{4}D^2 = \frac{\pi}{4}(2)^2 = \pi \, \text{m}^2$$
$$P = \pi D = 2\pi \, \text{m}$$
$$R = \frac{A}{P} = 0.5 \, \text{m}$$

Neglecting local losses, $S = h_L/L = 6/5000$. Hence

$$V = 0.849(120)(0.5)^{0.63}\left(\frac{6}{5000}\right)^{0.54} = 1.74 \, \text{m/s}$$

and

$$Q = AV = \pi(1.74) = 5.47 \, \text{m}^3/\text{s}$$

12. **b.** The Hazen-Williams equation may be written as

$$V = 0.849CAR^{0.63}S^{0.54}$$

or as

$$AR^{0.63} = \frac{Q}{0.849CS^{0.54}} = \frac{4.5}{0.849(130)\left[\dfrac{7.5}{6000}\right]^{0.54}} = 1.507$$

In this case

$$A = \frac{\pi}{4}D^2 \quad \text{and} \quad R = \frac{A}{P} = \frac{D}{4}$$

for a pipe flowing full. Thus

$$\frac{\pi}{4}D^2\frac{D^{0.63}}{4^{0.63}} = 1.507$$
$$D^{2.63} = \frac{1.507(4^{1.63})}{\pi} = 4.59$$

and

$$D = (4.59)^{\frac{1}{2.63}} = (4.59)^{0.38} = 1.785 \, \text{m}$$

13. **c.** From AISC Section B2, the net area of the brace is $A_n = A_g - 2td_h = 2 \times 2.09 - 2 \times 0.3125 (0.875 + 0.125) = 3.56$ square inches. From AISC Equation (B3-1), the effective net area of the brace, $A_e = UA_n = 0.85 \times 3.56 = 3.02$ square inches.

14. **d.** In accordance with AISC Section A5.2 a one-third increase in allowable stress is permitted for wind loading. The strength of the brace based on the effective net area, in accordance with AISC Section D1, is $P_t = 1.33 \times 0.5 \, F_uA_e = 1.33 \times 0.5 \times 58 \times 3.02 = 116.48$ kips.

15. c. The relevant properties of the W14 × 48 are $A_s = 14.1$ inches^2, $r_x = 5.85$ inches, and $r_y = 1.91$ inches. The slenderness ratio about the x-axis is given by $Kl/r_x = 2.1 \times 11 \times 12/5.85 = 47.4$. The slenderness ratio about the y-axis is given by $Kl/r_y = 0.8 \times 11 \times 12/1.91 = 55.3$, which governs. $C_c = (2\pi^2 \times 29,000/36)^{0.5} = 126.1 > Kl/ry$. The allowable stress is $F_a = [1 - 55.3^2/(2 \times 126.1^2)]36/[5/3 + 3 \times 55/3/(8 \times 126.1) - 55.3^3/(8 \times 126.1^3)] = 17.88$ kips per square inch. The allowable axial load is $P = 14.1 \times 17.88 = 252.11$ kips.

16. d. For a spirally reinforced column, the design axial load strength is given by the equation as $\phi P_n = 0.85\ \phi[0.85\ f'_c(A_g - A_{st})f_y A_{st}] = 0.85 \times 0.75[0.85 \times 35(196,350 - 7000) + 4000 \times 7000] = 5376$ kN.

17. b. In accordance with ACI Section 7.10.4.2, the minimum diameter of the spiral reinforcement is $\frac{3}{8}$-inch.

18. a. In accordance with ACI Section 10.91, the minimum reinforcement ratio required is 0.01 and the maximum allowable reinforcement ratio is 0.08.

19. c. The relationship between the external ordinate (E) and the radius (R) is:

$$E = \left[\frac{R}{\cos\frac{\Delta}{2}} - R \right]$$

The external ordinate of the existing curve is:

$$E = \left[\frac{300}{\cos\frac{40}{2}} - 300 \right] = 19.25 \text{ m}$$

The new curve is to be shifted radially a distance of 3 m. Therefore, the external ordinate of the new curve is 1925 − 3 = 16.25 m. The new radius is

$$R = \frac{E}{\frac{1}{\cos\frac{\Delta}{2}} - 1} = \frac{16.25}{\frac{1}{\cos\frac{40}{2}} - 1} = 253.20 \text{ m}$$

20. a. Determine the stopping sight distance for the 75-m-crest vertical curve.

$$L = (AS^2)/404$$
$$75 = (8S^2)/404$$
$$S = 61.54 \text{ m}$$

Determine the maximum safe speed for this sight distance.

$$S = 0.278Vt + V^2/[254\ (f \pm g)]$$
$$61.54 = 0.278(V)(2.5) + V^2/[254\ (0.40 - 0.04)]$$

From the quadratic equation:

$$V = 49.69 \text{ km/h}$$

21. b.

$$\text{PHF} = V_h/(4V_{15})$$

where V_h = hourly volume (vph), V_{15} = maximum 15-minute rate of flow within the hour (veh).

$$\text{PHF} = (1000 + 1100 + 1200 + 900)/(4 \times 1200) = 0.875$$

22. c.

$$\text{Peak flow rate} = 4 \times 1200 = 4800 \text{ veh/hr}$$

23. d. The relationship between speed and length of spiral is:

$$\lambda_s = V^3/(46.7RC)$$
$$52.083 = V^3/[(46.7)(300)(0.6)]$$
$$V = 75.9 \text{ km/h}$$

24. a. Compute the spiral tangent.

$$T_s = (R + p) \tan (\Delta/2) + k = (300 + 0.377) \tan (16/2) + 26.035 = 68.250 \text{ m}$$
$$\text{TS station} = \text{PI station} - T_s = 1 + 086.271 - 0 + 068.250 = 1 + 018.021$$

25. b. Stopping sight distance (S) is

$$S = 0.278Vt + V^2/[254(f \pm g)]$$
$$= 0.278(80)(2.5) + (80)^2/[254(0.35 + 0.03)] = 121.9 \text{ m}$$

26. d.

Station	Area, m^2	Total Area	Average Area	Distance, m	Volume[a], m^3
5 + 000	27.9				
		354	177.0	50	8850.0
5 + 050	326.1				
		1291.4	645.7	50	32,285.0
5 + 100	965.3				
				Total Volume	= 41,135.0

[a] Volume = distance × average area

27. a. The bases of the cross sections are constant (11 m). The heights of the cross sections vary between stations but can be determined by dividing the cross section areas by 11 m. The prismoidal formula is

$$V = (L/6)(A_1 + 4A_m + A_2)$$

Determine the volume between stations 1 + 130 and 1 + 200.

$$A_m = [(3.2 + 3.6)/2]11 = 37.4 \text{ m}^2$$
$$V = (70/6)[35 + 4(37.4) + 40] = 2620.3 \text{ m}^3$$

Determine the volume between stations 1 + 200 and 1 + 260.

$$A_m = [(3.6 + 4.1)/2]11 = 42.4 \text{ m}^2.$$
$$V = (60/6)[40 + 4(42.4) + 45 = 2546.0 \text{ m}^3]$$

Total volume = $2620.3 + 2546.0 = 5166.3 \text{ m}^3$

28. c. The binary (base 2) number system representation is

2^4	2^3	2^2	2^1	2^0
16	8	4	2	1

Binary number	1	0	1	0	1

16	0	4	0	1 = 21

29. d.

2^5	2^4	2^3	2^2	2^1	2^0
32	16	8	4	2	1
	1	1	1	1	0 = 30

30. b.

31. b.

32. b. A cell reference is made by a lettered column and a numbered row, for example, C3.

33. b. Synthetic division

1	4	6	4	−1	trial root
	−1	−3	−3		
1	3	3	1		a remainder of 1

34. d.

$$x^{n+1} = x_n - P(x_n)/P'(x_n)$$

for $x_1 = -1$, $P'(x) = 3x^2 + 8x + 6 \mid_{x = -1} = 3 - 8 + 6 = 1$

$$P(x)\mid_{x = -1} = -1 + 4 - 6 + 4 = 1$$
$$\therefore x_{n+1} = (-1) - (1/1) = -2$$

35. c. For fixed increments

$$(x_{k+1}) = x_k - Tax_k + Tf_k$$
$$x_1 = x_0 - Ta_{x0} + Tf = 0 - 0 + 0.3 \times 1 = 0.3$$

36. c. Choices a, b, and d represent practical and appropriate ways of improving one's competence. Choice c however may subject the client or employer to work by an engineer who is not fully competent in the specific area. This is clearly a less appropriate way to gain competency in a technical area.

37. d. Actions a, b, and c are prescribed in the *Model Rules*. Section 240.15(I)(e) states that engineers shall express a professional opinion publicly only when it is based on adequate knowledge and a competent evaluation. Choice d is commenting publicly *before* a careful evaluation.

38. d. Professional conduct precludes an engineer from doing the things in a, b, and c. Section 240.15(II)(f) states that the engineer shall not accept compensation from more than one party for services on the same project unless the circumstances are fully disclosed and agreed to by all parties. Thus the conduct in d is acceptable.

39. c. The best answer is c. If a is chosen and discovered, you may jeopardize your employment. Working at the office and using company resources, tools, or proprietary methods would be unethical, so answer b must be excluded. Answer d might be ethically acceptable but relies on involved parties as witnesses to an agreement.

40. c. As a professional engineer you have a responsibility to safeguard life, health, and property. Thus you should take appropriate action to advise the proper person or people of the need to review the fuse selection. The question becomes who should one contact. While alternative b might be all that is needed, alternative c is a more formal action (memo instead of a call) and is addressed to the supervisor level, and is therefore preferred. Alternative d is premature since you do not know the current state of the design (the fuse size may already have been changed). Almost certainly a is not acceptable to your employer or in accordance with his policies on how to handle matters like this.

41. c. Although, in a legal sense, only action a is unacceptable; items b and d may possibly not be in the best interest of everyone because of unknown human effects, therefore item c is in the best interest of public safety.

42. c.

43. a.

44. b.

45. a.

46. **d.** All of the statemeents are correct.

47. **a.**

48. **b.**

$$Q_f = 0.2 \text{ m}^3/\text{s}$$
$$Q/Q_f = 0.4$$
$$Q = 0.08 \text{ m}^3/\text{s}$$

49. **c.**

50. **a.**

$$\frac{A}{A_0} = \frac{1}{1 + K_1 \dfrac{v}{Q}}$$

$$0.2 = \frac{1}{1 + K_1 \dfrac{v}{Q}}$$

$$1 + K_1 \frac{v}{Q} = 5$$

$$v = \frac{4Q}{K_1} = \frac{3.46 \times 10^5}{0.3} = 1.1 \times 10^6 \text{ m}^3$$

51. **a.**

52. **d.**

53. **c.**

54. **c.** Crane tips about A when $R_B = 0$

$$\sum M_A = -36\,W + 100 \text{ kips } (12) = 0$$
$$W = \frac{100}{3} = 33.3 \text{ kips}$$

If $W > 33.3$, crane will tip about A.
Crane tips about B when $R_A = 0$

$$\sum M_B = -W(44) + 100(4) = 0$$
$$W = \frac{400}{44} = 9.1 \text{ kips}$$

Hence W must be between 9.1 and 33.3 kips.

55. **b.** Use FBD shown

$$\sum V = 0 \uparrow +$$
$$-5 - 10 + 0.707(\text{EF}) = 0$$
$$\text{EF} = +21.2$$

Assumed direction ok, so,

$$EF = 21.2 \text{ kips tensile}$$

56. c.

$$\sum H = 0 \quad \text{gives} \quad R_B \text{ (horiz comp)} = \leftarrow 12 \text{ kips}$$

$$\sum N_A = 0 \quad \text{gives} \quad R_B \text{ (vert comp)} = \downarrow 19.6 \text{ kips}$$

$$\text{Resultant} \approx \downarrow 23 \text{ kips}$$

57. b. From $\Sigma H = 0$, $H_A = 0$
$\Sigma V = 0 \downarrow +$ gives (FBD of beam AB)

$$+28.57 - 2 \times 10 + V_A(\uparrow) = 0$$
$$V_A (\uparrow) = 8.57$$

58. b. FBD whole structure.

$$\sum H: \quad H_A = 0 \text{ kip-ft ccw.}$$

$$\sum V: \quad V_A = 5 \text{ kip-ft ccw.}$$

$$\sum M: \quad M_A = 70 \text{ kip-ft ccw.}$$

ΣM_A gives $v = 70/9 = 7.77$
Since strut is @45°, $v = h$
Hence shear at A–A = $7.77 - 5.00 \cong 2.8$ kips.

59. b.

$$\sum M_{BB} \text{ clockwise} = +5(.7) - 7.77(2) + M_{BB} = 0$$
$$M_{BB} = -(35 - 15.54) = -19.4 \text{ kip-ft}$$

60. b. Find the flow depth, d, in feet. Manning's formula

$$v = \frac{1}{n} R^{\frac{2}{3}} S^{\frac{1}{2}}$$

where
   v = velocity
   $R = A/P$
   A = channel cross-sectional area
   P = wetted perimeter
gives

$$Q = AV = \frac{1}{n} A R^{\frac{2}{3}} S^{\frac{1}{2}}$$

For this trapezoidal channel

$$A = (1.5 + 2d)d \quad \text{and} \quad P = 1.5 + 2\sqrt{5}d$$

Hence

$$15 = \frac{1}{0.015}(1.5+2d)d\left[\frac{(1.5+2d)d}{1.5+2\sqrt{5}d}\right]^{\frac{2}{3}}(0.0034)^{\frac{1}{2}}$$

Solve by trial for d; that is, choose a value of d, substitute into the equation and see whether it is satisfied. An example of the progress of this solution starting with $d = 1.0$ is shown in the table below.

$$AR^{2/3} = 3.8587$$

where
$A = (1.5 + 2d)d$
$P = 1.5 + 2\sqrt{5}d$
$R = A/P$

d	A	P	R	$R^{\frac{2}{3}}$	$AR^{\frac{2}{3}} \overset{?}{=} 3.8587$
1.0	3.50	5.97	0.586	0.701	2.45
1.5	6.75	8.21	0.822	0.878	5.92
1.2	4.68	6.87	0.681	0.774	3.62
1.25	5.00	7.09	0.705	0.792	3.96
1.23	4.87	7.00	0.696	0.785	3.82

We find $d = 1.23$ m, which gives $A = 4.87$ m^2.